Regional Water Security

Regional Water Security

Robert C. Brears

Christchurch, New Zealand

WILEY Blackwell

This edition first published 2021

Registered Offices
John Wiley & Sons, Inc., 111 River Street, Hoboken, NJ 07030, USA
John Wiley & Sons Ltd, The Atrium, Southern Gate, Chichester, West Sussex, PO19 8SQ, UK

Editorial Office
9600 Garsington Road, Oxford, OX4 2DQ, UK

For details of our global editorial offices, customer services, and more information about Wiley products visit us at www.wiley.com.

Wiley also publishes its books in a variety of electronic formats and by print-on-demand. Some content that appears in standard print versions of this book may not be available in other formats.

Library of Congress Cataloging-in-Publication Data

Names: Brears, Robert C., author.
Title: Regional water security / Robert C. Brears, Christchurch, New Zealand.
Description: Hoboken, NJ, USA : Wiley-Blackwell, 2021. | Includes bibliographical references and index.
Identifiers: LCCN 2020040377 (print) | LCCN 2020040378 (ebook) | ISBN 9781119661139 (cloth) | ISBN 9781119661115 (adobe pdf) | ISBN 9781119661122 (epub)
Subjects: LCSH: Water-supply. | Water security–Government policy. | Water quality management.
Classification: LCC TD345 .B75 2021 (print) | LCC TD345 (ebook) | DDC 333.91–dc23
LC record available at https://lccn.loc.gov/2020040377
LC ebook record available at https://lccn.loc.gov/2020040378

Cover Design: Wiley
Cover Images: Inna Bigun/Shutterstock

Set in 9.5/12.5pt STIXTwoText by SPi Global, Pondicherry, India
Printed and bound by CPI Group (UK) Ltd, Croydon, CR0 4YY

C9781119661139_030421

Contents

Acknowledgments

First, I wish to first thank Andrew Harrison for being a visionary who enables books like mine to come to fruition. Second, I wish to thank Mum, who has a great interest in the environment and has supported me in this journey of writing the book.

1

Introduction

Regional water security around the world is at risk from both climatic and nonclimatic challenges impacting water quantity and water quality. For instance, climate change is projected to decrease the availability of renewable surface water and groundwater resources significantly, intensifying competition for water resources among users, as well as to reduce the availability of good-quality drinking water, even with conventional treatment processes (Arnell et al. 2015; Brears 2016a). Rapid population growth and urbanization will lead to increased water scarcity and ecosystem degradation due to excess water withdrawal, eutrophication, land-use changes, and pollution (Brears 2016a, 2018), while peri-urban water competition is likely to increase with further urbanization (McDonald et al. 2014). With economic growth, global demand for water will significantly increase due to manufacturing, industry, and domestic consumption (PwC 2015; UN Water 2014). At the same time, household water demand is projected to increase due to higher incomes and living standards, as well as a shift in diet to water-intensive meat and dairy products (Harlan et al. 2009; Kearney 2010; UNESCO 2012; UN Water 2014). In many locations, a large portion of the water infrastructure is approaching or has already reached the end of its useful life, with aging infrastructure often resulting in high water loss from physical leakage (Lam et al. 2017). In addition, sewage and contaminated groundwater can enter leaking pipes and travel throughout the water distribution network, causing public health concerns, such as outbreaks of gastrointestinal illness (Fox et al. 2016; Säve-Söderbergh et al. 2017). Between now and 2040, it is projected that the amount of energy used in the water sector will double, with the most substantial increase coming from desalination, followed by large-scale water transfer and increasing demand for (higher levels of) wastewater treatment (IEA 2016). Finally, by 2050, the world will require at the minimum 60 percent more food production to maintain current consumption patterns, leading to a significant increase in the volume of global water withdrawn for irrigation. Agriculture is one of the most significant contributors to non-point-source pollution impacting rivers, streams, and lakes – as well as wetlands and groundwater supplies – harming public health, creating dead zones in water, increasing water treatment costs, and impacting industries dependent on clean water (FAO 2015; Brears 2016b).

Regional development, which can be defined as community development (social, economic, environmental, technological, cultural, and recreational) of a particular territory, has proved challenging in practice as sustainable development at the regional level implies

Regional Water Security, First Edition. Robert C. Brears.

the implementation of complementary and coordinated actions in different areas, resulting in economic growth that also supports social and environmental objectives (Jovovic et al. 2017). In the context of water resource management, managing water in regions to achieve various economic, environmental, and social goals is particularly challenging given that the majority of the world's water resources are transboundary, crossing both intra- and interstate administrative and political lines.

In response to decreasing water quantity and diminishing water quality, water authorities, at multiple levels of governance around the world, are implementing a variety of policies to encourage the transition toward water security for both humans and nature, focusing on balancing rising demand with limited – and often variable – supplies of water and developing green infrastructure to mitigate climatic risks and improve water quality, all the while realizing multiple regional development goals.

Traditionally, water managers have relied on large-scale, supply-side infrastructural projects such as dams and reservoirs to meet increased demands for water (supply-side management). However, these projects are costly, both economically and environmentally. Also, with most water resources being transboundary, supply-side projects can create political tensions. Therefore, there is a need to move toward managing actual demand for water (demand management), as ultimately it is society's attitudes and behaviors toward water that determine the amount of that needs to be supplied. Demand management involves making better use of existing water supplies before attempting to increase them further. Specifically, it promotes water conservation under both normal and abnormal conditions, through changes in practice, culture, and people's attitudes toward water resources. It seeks to reduce the loss and misuse of water, optimize its use, and facilitate major financial and infrastructural savings by minimizing the need to meet increasing demand with new water supplies. Demand management also involves the development of alternative water supplies as a means of diversification in order to meet various potable and nonpotable water needs (Brears 2016a).

The most common means of mitigating risks from climatic extremes has been increasing investment in conventional – or "gray" – infrastructures such as dams and levees. However, engineers and decision-makers have come to realize the economic and environmental costs of these solutions, including high capital costs, amplified downstream flooding risks, and ecosystem degradation. As such, there has been a turn to more long-term economically and environmentally sustainable "green" infrastructure solutions that provide equivalent or similar benefits to gray infrastructure. For instance, gray infrastructure utilizes natural and semi-natural systems to manage excess water while enhancing ecosystems. It can also be used to mitigate the impacts of droughts, ensuring there is sufficient water for both humans and nature. Furthermore, gray infrastructure can purify water using natural filtration processes. A unique feature of gray infrastructure is that it can appreciate over time (Brears 2018).

Nonetheless, our understanding of the role regional water managers have in implementing demand management strategies and green infrastructure to enhance water security and reduce environmental degradation lags significantly behind our engineering knowledge of water resources management. As such, little has been written on the actual implementation of green policy innovations across entire regions that not only ensure water security for humans but also restore ecosystems and the numerous services they provide. Also, because the application of demand management and green infrastructure requires holistic planning, little has been written on how innovative policies have been developed to ensure they fulfil multiple functions and policy goals and create multiple benefits for society and ecosystems.

Regional Water Security provides new research on policy innovations that promote the application of demand management and green infrastructure in managing water resources across regions sustainably. In particular, it offers in-depth case studies that illustrate how different regions – of differing climates, lifestyles, and income levels – are implementing policy innovations which promote the application of demand management and green infrastructure to achieve regional water security for humans while protecting and restoring the natural environment. A summary runs as follows:

Chapter 2: Water Security This chapter discusses the concept of water security and defines what it means at the regional level. It then looks at the various climatic and nonclimatic challenges to regional water security.

Part I: Demand Management

Chapter 3: Water Allocation This chapter discusses how water allocation involves implementing technologies and establishing arrangements for sharing existing resources among users equitably and to the maximum benefit of all, including the environment. Technologies discussed (with examples of their implementation around the world) include basin-level modelling and seasonal forecasting, seasonal water rationing, and water reallocation.

Chapter 4: Water Augmentation This chapter discusses how water augmentation aims to increase the available supply of water through active recharge or protection of water recharge areas. It provides global examples of technologies being implemented, including the combined use of surface and groundwater, aquifer recharge, and source water protection.

Chapter 5: Water Efficiency This chapter discusses how residential, irrigation, and industrial water efficiency can be increased through more effective leak detection and repair of water pipes, improved irrigation efficiency to minimize water use within the agricultural sector while continuing to maintain crop productivity rates, and the use of new and more efficient technologies in industrial facilities.

Chapter 6: Water Reuse and Water Recycling This chapter discusses a range of water reuse and water recycling technologies and initiatives that have been implemented around the world, including rainwater and stormwater harvesting systems, potable and nonpotable reuse systems, and other types of alternative water source technologies.

Part II: Green Infrastructure

Chapter 7: Green Buildings and Green Streets This chapter reviews the various types of green infrastructure implemented at the building and street level. In particular, it looks at the use of green roofs, blue roofs, rain barrels, rainwater harvesting systems, and green streets with permeable pavement and underground systems to capture and purify stormwater.

Chapter 8: Green Parks and Urban Forests This chapter discusses how green parks and urban forests intercept and filter stormwater runoff, preventing flooding and improving water quality, while at the same time providing multiple co-benefits.

Chapter 9: Water Bodies This chapter discusses various green infrastructure strategies that can be implemented to decrease flood risk and improve water quality while protecting, conserving, and enhancing natural ecosystems, including retention ponds and riparian buffers, as well as river engineering, floodplain restoration, wetland restoration, and green corridor initiatives.

Chapter 10: Agriculture and Forestry This chapter discusses how natural water retention measures, which are a form of green infrastructure, can play an essential role in catchment-scale flood risk management by enhancing or restoring natural hydrological processes, including interception, evapotranspiration, infiltration, and ponding.

Part III: Case Studies, Best Practices, and Conclusions

Chapter 11: Case Studies of Regions Implementing Demand Management and Green Infrastructure to Achieve Regional Water Security This chapter provides case studies of leading locations around the world that are implementing a variety of demand management and green infrastructure initiatives to achieve regional water security.

Chapter 12: Best Practices This chapter discusses a range of best practices identified in the application of demand management and green infrastructure that can be implemented by regions attempting to achieve regional water security for humans while at the same time protecting and restoring the natural environment.

Chapter 13: Conclusions

References

Arnell, N.W., S.J. Halliday, R.W. Battarbee, R.A. Skeffington, and A.J. Wade. (2015). "The Implications of Climate Change for the Water Environment in England." *Progress in Physical Geography: Earth and Environment* 39, no. 1: 93–120.

Brears, R.C. (2016a). *Urban Water Security*. Hoboken, NJ: John Wiley & Sons.

Brears, R.C. (2016b). "The Water–Food Nexus and the Role of Demand Management." *Future of Food: Journal on Food, Agriculture and Society* 4, no. 3: 73–74.

Brears, R.C. (2018). *Blue and Green Cities: The Role of Blue-Green Infrastructure in Managing Urban Water Resources*. London: Palgrave Macmillan.

FAO. (2015). "Towards a Water and Food Secure Future: Critical Perspectives for Policy-Makers." http://www.fao.org/3/a-i4560e.pdf (accessed October 9, 2020).

Fox, S., W. Shepherd, R. Collins, and J. Boxall. (2016). "Experimental Quantification of Contaminant Ingress into a Buried Leaking Pipe During Transient Events." *Journal of Hydraulic Engineering* 142, no. 1: 04015036.

Harlan, S.L., S.T. Yabiku, L. Larsen, and A.J. Brazel. (2009). "Household Water Consumption in an Arid City: Affluence, Affordance, and Attitudes." *Society & Natural Resources* 22, no. 8: 691–709.

IEA. (2016). "WEO-2016 Special Report: Water–Energy Nexus." https://webstore.iea.org/weo-2016-special-report-water-energy-nexus (accessed October 9, 2020).

Jovovic, R., Draskovic, M., Delibasic, M., and Jovovic, M. (2017). "The Concept of Sustainable Regional Development – Institutional Aspects, Policies and Prospects." *Journal of International Studies* 10, no. 1.

Kearney, J. (2010). "Food Consumption Trends and Drivers." *Philosophical Transactions of the Royal Society of London. Series B, Biological Sciences* 365, no. 1554: 2793–2807.

Lam, K.L., S.J. Kenway, and P.A. Lant. (2017). "Energy Use for Water Provision in Cities." *Journal of Cleaner Production* 143: 699–709.

McDonald, R.I., K. Weber, J. Padowski, M. Flörke, C. Schneider, P.A. Green, et al. (2014). "Water on an Urban Planet: Urbanization and the Reach of Urban Water Infrastructure." *Global Environmental Change* 27: 96–105.

PwC. (2015). "The World in 2050: Will the Shift in Global Economic Power Continue?" http://www.pwc.com/gx/en/issues/the-economy/assets/world-in-2050-february-2015.pdf (accessed October 9, 2020).

Säve-Söderbergh, M., J. Bylund, A. Malm, M. Simonsson, and J. Toljander. (2017). "Gastrointestinal Illness Linked to Incidents in Drinking Water Distribution Networks in Sweden." *Water Research* 122: 503–511.

UNESCO. (2012). "Managing Water Under Uncertainty and Risk." http://www.unesco.org/new/fileadmin/MULTIMEDIA/HQ/SC/pdf/WWDR4%20Volume%201-Managing%20Water%20under%20Uncertainty%20and%20Risk.pdf (accessed October 9, 2020).

UN Water. (2014). "Partnerships for Improving Water and Energy Access, Efficiency and Sustainability." http://www.un.org/waterforlifedecade/water_and_energy_2014/pdf/water_and_energy_2014_final_report.pdf (accessed October 9, 2020).

2

Water Security

Introduction

The United Nations defines water security as "the capacity of a population to safeguard sustainable access to adequate quantities of acceptable quality water for sustaining livelihoods, human well-being, and socio-economic development, for ensuring protection against water-borne pollution and water-related disasters, and for preserving ecosystems in a climate of peace and political stability" (UN Water 2013). Societies can achieve water security when they: successfully manage their water resources and services to satisfy household water and sanitation needs in all communities; support productive economies in agriculture, industry, and energy; develop vibrant, livable cities and towns; restore healthy rivers and ecosystems; and build resilient communities that can adapt to change. In this context, this chapter will discuss the main dimensions to water security and the trends that impact its attainment, before going on to look at the role of demand management and green infrastructure in this area.

Dimensions to Water Security

There are five critical dimensions to achieving water security:

1) *Household Water Security*: The foundation and cornerstone of water security is what happens at the household level. Providing all people with reliable, safe water and sanitation services should be the top priority in all locations. Household water security is essential to eradicating poverty and supporting economic development.
2) *Economic Water Security*: Water is used for agricultural production, in industry, and to cool energy-generating plants. These uses must not be viewed in isolation from each other. The water–energy–food nexus raises awareness of the critical interaction of water uses in supporting economic activities. Economic water security measures the productive use of water to sustain economic growth in the food production, industry, and energy sectors.
3) *Urban Water Security*: The world is rapidly urbanizing, with cities becoming the drivers of the global economy. As such, urban water security is achieved through better water management and services in support of vibrant and livable water-secure cities.

Regional Water Security, First Edition. Robert C. Brears.

4) *Environmental Water Security*: Commonly, regions around the world have prioritized economic growth over ecological objectives. There is now a shift toward green growth that focuses on broader sustainable development. Environmental water security is achieved through the restoration of rivers and ecosystems on the national and regional scales. The sustainability of development and improved lives depends on these natural resources.
5) *Resilience to Water-Related Disasters*: The building of resilient communities that can adapt to change and reduce risk from natural disasters related to water must be accelerated in order to minimize the impact of future disasters (ADB 2016).

Addressing Water Security Risks

Achieving water security means maintaining acceptable levels for four water risks:

1) *Risk of Shortage (Including Droughts)*: Lack of sufficient water to meet demand – in both the short and the long term – from all water users (households, businesses, and the environment).
2) *Risk of Inadequate Quality*: Lack of water of adequate quality for a particular purpose or use.
3) *Risk of Excess (Including Floods)*: Overflow of the normal confines of a water system (natural or built), or the destructive accumulation of water over areas that usually are not submerged.
4) *Risk of Undermining the Resilience of Freshwater Systems*: Exceedance of the coping capacity of surface and groundwater bodies and their interactions (the "system") (possibly crossing tipping points and causing irreversible damage to the system's hydraulic and biological functions).

All four risks need to be assessed in an integrated way, as interventions to reduce one can lead to increases in others. As such, a risk-based approach to water security is recommended. In particular, it is important to:

1) *Know the Risks*: A significant obstacle to water security is often a lack of knowledge about water risks and their scale. It is essential to understand such risks and anticipate and address all potential risk drivers, including socioeconomic trends, natural phenomena, and inadequate water and water-related policies.
2) *Target the Risks*: The acceptable level of water risk for a society should depend on the balance between economic, social, and environmental consequences and the cost of amelioration. However, eliminating risk is often technically impossible or prohibitively expensive.
3) *Manage the Risks*: Allocating water risks between residential, agricultural, industrial, and environmental uses raises significant questions. A risk-based approach allows for the assignment of risks to the actors who are likely to be able to manage them most efficiently. A risk-based approach also helps ensure an equitable distribution of risks among stakeholders and can help prevent the imposition of one group's risk preferences on others (OECD 2016).

Changes in Water Quantity and Water Quality Affect Water Security

Changes in water quantity and water quality and associated societal interactions impact the ability to achieve water security. Table 2.1 outlines the effects of variations on water quality (Gunda et al. 2019; OECD n.d.), while the impacts of changes to water quantity are as follows:

- Reduced stream habitats.
- Depletion of soil moisture, with knock-on effects on plant growth, due to elevated rates of evaporation.
- Loss of or reduction in crop yields and production.

Table 2.1 Impacts of variable water quality.

Area of Impact	Examples
Urban and domestic	Increased water treatment and inspection costs Maintenance costs from scouring and premature aging of infrastructure Increased wastewater treatment costs with the implementation of more strict regulations Emergency and clean-up costs from spills/accidents
Ecosystem health	Damage to freshwater and marine ecosystems (e.g., fish kill, habitat destruction, and loss of ecosystem services), which may require investment in additional or different gray infrastructure alternatives to replicate these services
Human health	Polluted water – the world's most significant health risk, affecting the quality of life and public health; health service costs, loss of life expectancy, and emerging health costs are all associated with pollution events
Industrial productivity	Exclusion of contaminated water for industrial use, resulting in increased water scarcity and lower industrial productivity Scouring of infrastructure and clean-up costs from spills/accidents
Social values and tourism	Prohibition on recreational use (e.g., swimming, fishing, seafood gathering) Beach closure Impacts on esthetic, cultural, and spiritual values Losses in fishing, boating, rafting, and swimming activities to other tourism activities or other ventures with superior water quality
Agricultural productivity	Exclusion of contaminated water from irrigation, increasing water scarcity Irrigation with contaminated water, causing damage to – and reduced productivity of – pasture and crops Soil contamination, impacting livestock health and production
Commercial fisheries	Direct and indirect fish kill Contamination of shellfish
Property values	Decline in waterfront property values due to unsightly pollution and odor

Source: Thushara Gunda et al., "Water Security in Practice: The Quantity-Quality-Society Nexus," Water Security 6(2019); OECD, "Infographic – What Are the Impacts of Water Pollution?," https://www.oecd.org/water/infographic-impacts-of-water-pollution.htm.

- Minimal grass cover and loss of income from selling off stock.
- Loss of production in various water-using industries.
- Little recharge of groundwater, resulting in low spring flows (baseflows to rivers) and low groundwater levels.
- Low levels in major storage dams and lakes, and drying up of shallow bore springs and streams.
- Costly transport of water to rural residents.
- Restrictions on the use of water imposed by water authorities.
- Loss of tourism due to water use bans and water shortages.
- Decreased energy production due to higher river water temperatures, limiting the water's use as a coolant, and water reservoirs dropping below an efficient level for hydropower facilities.
- Intensification of pollution by nutrients and pesticides as there is less water to dilute excess nutrients and toxic substances that have leached into water bodies, negatively affecting fish spawning and increasing algal blooms.
- Water-related diseases resulting from drinking insufficient amounts of water or water with high salinity (Hunter et al. 2010; Northland Regional Council 2011; Pierre Strosser et al. 2012; Abedin et al. 2019).

Trends Impacting Water Security

There are many trends impacting water security globally, including the following.

Climate Change

Climate change is likely to lead to shifts in precipitation patterns and snow cover, increasing the frequency of flooding and droughts. Higher temperatures will generally intensify the global hydrological cycle. In some parts of the world, annual precipitation levels will increase, while in other regions they will decrease. With snow limits pushed upward and less precipitation falling as snow, many locations will experience higher winter runoff and low summer flow rates. Climate change will increase the frequency and intensity of rainfall: for every 1 °C increase in temperature, the air can hold about 7% extra water vapor. In some locations, there may be an increase in the occurrence of flooding due to heavy rainfall events. Globally, it is estimated that the number of people affected by flooding will rise from 65 million in 2010 to 132 million in 2030. Meanwhile, it is estimated that for each degree of global warming, 7% of the world's population is exposed to a decrease in renewable water resources of at least 20%. Already, more than 2 billion people live in countries experiencing high water stress. Some of the impacts of climate change on water resources are listed in Table 2.2 (IPCC 2014; US EPA 2017; EEA 2018; UN Water 2019a, n.d.; World Resources Institute 2020).

Water-Related Disasters

Water-related disasters account for 90% of the 1000 most severe disasters that have occurred since 1990. There are three primary hazards (floods, droughts, and extreme storms) that can lead to water-related disasters affecting people or economic goods, where there is insufficient preparedness.

Table 2.2 Impacts of climate change on water resources.

Impact	Description
Increased droughts	Climate change is likely to increase the frequency of meteorological droughts (less rainfall) and agricultural droughts (less soil moisture) in many presently dry regions
Lower water quality	Climate change will alter streamflow and water quality Increase in heavy precipitation events can lead to sewer systems and wastewater treatment plants being overwhelmed by the increased volumes of water Heavy downpours can increase the amount of runoff into waterways, washing sediment, nutrients, pollutants, rubbish, animal waste, and other materials into water supplies, making them unusable, unsafe, or in need of water treatment
Lower snowmelt	In regions with snowfall, climate change has altered – and is likely to continue altering – streamflow seasonality Reduced spring maximum snow depth and earlier spring snowmelt are leaving less snow in storage for dry summer months
Rising sea levels	Freshwater resources along coastal regions face risks from sea level rise Rising sea levels and the occurrence of drought can increase the salinity of both surface water and groundwater through saltwater intrusion

Source: UN-Water, "Un-Water Policy Brief on Climate Change and Water," (2019); IPCC, "The Fifth Assessment Report (Ar5) of the Intergovernmental Panel on Climate Change," (2014), https://www.ipcc.ch/assessment-report/ar5/; US EPA, "Climate Impacts on Water Resources," https://19january2017snapshot.epa.gov/climate-impacts/climate-impacts-water-resources_.html; US EPA, "Climate Impacts on Water Resources," https://19january2017snapshot.epa.gov/climate-impacts/climate-impacts-water-resources_.html; World Resources Institute, "The Number of People Affected by Floods Will Double between 2010 and 2030," https://www.wri.org/blog/2020/04/aqueduct-floods-investment-green-gray-infrastructure.

Floods affect the highest number of people annually: over the period 1995–2015, they affected 2.3 billion, killed 157 000, and caused $662 billion in damage. Flooding can be classified into urban flooding, pluvial flooding (flash flooding), fluvial flooding (exceedance of river channel capacity, resulting in floodplain inundation), waterlogging, and coastal flooding. The immediate impacts include loss of human life, damage to property, destruction of crops, loss of livestock, and deterioration of health conditions due to waterborne diseases. Floods impact economic activities by damaging and disrupting communication links and infrastructure, including power plants, roads, and bridges. Meanwhile, disruption to industry can lead to loss of livelihood. Damage to infrastructure may have long-term impacts, such as interruptions to supplies of clean water, wastewater treatment, education, and healthcare. Increases in population growth and changes in land-use patterns have increased human vulnerability to floods (Doocy et al. 2013; Office of the Queensland Chief Scientist 2018).

Droughts accounted for 5% of all natural disasters between 1995 and 2015, affecting 1.1 billion people, killing 22 000, and causing $100 billion in damage. They can impact the environment and communities, with impacts grouped as "economic," "environmental," and "social." Some of their effects are listed in Table 2.3 (Stanke et al. 2013; NCEH 2016; Kuwayama et al. 2018; UN Water 2019b; NDMC n.d.).

Table 2.3 Impacts of droughts.

Impact	Examples
Economic	Loss of crops
	Increased irrigation or well development costs
	Losses among rural-dependent businesses due to damage to crops or livestock
	Losses in the timber industry due to increased wildfires
	Increased electricity costs due to lower hydroelectric output
	Difficulty for barges and ships in navigating waterways, affecting water transportation businesses
	Costs to water utilities due to the need for new or additional water supplies
Environmental	Losses or destruction of fish and wildlife habitats
	Lack of food and drinking water for animals
	Increases in disease in animals due to reduced food and water supplies
	Migration of wildlife
	Increased stress on endangered species, or even extinction
	Lower water levels in reservoirs, lakes, and ponds
	Loss of wetlands
	More wildfires
	Wind and water erosion of soils
	Poor soil quality
Social	Anxiety or depression about economic losses
	Health problems related to low water flows and poor water quality
	Health problems related to dust
	Loss of human life
	Threat to public safety from increased number of forest fires
	Reduced incomes
	Migration of people from farms into cities or from one city to another
	Fewer recreational activities
	Lower water supplies for at-risk populations such as people in hospitals

Source: UN Water, "World Water Development Report 2019: Leaving No One Behind," (2019), https://knowledge.unccd.int/publications/world-water-development-report-2019-leaving-no-one-behind; Carla Stanke et al., "Health Effects of Drought: A Systematic Review of the Evidence," PLoS currents 5(2013); National Drought Mitigation Center, "How Does Drought Affect Our Lives?," https://drought.unl.edu/Education/DroughtforKids/DroughtEffects.aspx; National Center for Environmental Health, "Drought and Public Health in the U.S," (2016), https://www.cdc.gov/nceh/hsb/cwh/hsb:drought_infographic508-cleared.pdf; Yusuke Kuwayama et al., "Estimating the Impact of Drought on Agriculture Using the U.S. Drought Monitor," American Journal of Agricultural Economics 101, no. 1 (2018).

Storms, including hurricanes, cyclones, and storm surges, killed more than 242 000 people over the period 1995–2015, making them the deadliest type of weather-related disaster over the past two and a half decades. Storms can cause significant damage to water and wastewater treatment plants, as well as to buried wastewater and stormwater infrastructure, with common types of damage including the separation of pipe joints, leaks, and breaks (Chisolm and Matthews 2012; ADB 2015; UNISDR 2015).

Population Growth and Urbanization

By 2050, the world's population is expected to reach 9.7 billion, up from 7.7 billion in 2019. Nine countries will make up more than half of the projected growth of the global population between now and mid-century: India, Nigeria, Pakistan, the Democratic Republic of Congo, Ethiopia, the United Republic of Tanzania, Indonesia, Egypt, and the United States of America. The population of sub-Saharan Africa is projected to double by 2050, while other regions will experience lower rates of growth (e.g., 28% in Australia and New Zealand and 2% in Europe and North America) (UN DESA 2019a). Access to clean water is a significant issue for the world's population, with 47% already living in areas that suffer water scarcity at least one month a year. By 2050, it is estimated that this figure will rise to 52%. Meanwhile, during the mid-2010s, 1.9 billion people lived in potentially water-scarce areas, but by 2050 this number will increase to 2.7–3.2 billion. Groundwater is under pressure from rising populations, with more than 30% of the world's largest groundwater systems already in distress (Boretti and Rosa 2019).

Globally, 55% of the world's population resided in urban areas in 2018, up from 30% in 1950. By 2050, it is projected that 68% of the population will be urban. The level of urbanization is expected to increase in all regions, with Latin America and the Caribbean and North America rising from over 80% urbanized in 2018 to nearly 90% in 2050, and Europe rising from 75% today to 80% by mid-century. Over the period 1950–2018, the urban population of the world rose from 751 million to 4.2 billion (UN DESA 2019b). Growth in the urban population is driven by overall population growth and by an increasing percentage of the population living in urban areas. These two factors are projected to add 2.5 billion people to the world's urban population by 2050, with 90% of this growth occurring in Africa and Asia. While nearly half of the world's urban population will reside in urban areas with fewer than 500 000 inhabitants, the number of mega-cities with more than 10 million inhabitants will increase from 33 today to 43 by 2030 (UN DESA 2019b).

By 2050, demand for water will have increased by 55%, mainly due to urbanization (IWA n.d.). The result will be increased intersectoral demand for water and rising urban/rural competition for scarce resources. Urbanization has already led to increased competition between cities and agriculture for water, with 69 cities – with a total population of 383 million people – currently receiving around 16 billion cubic meters of reallocated water per annum (nearly equivalent to the annual flow of the Colorado River). Meanwhile, 21 cities rely on multiple water reallocation projects, including Amman, Jordan and Hyderabad, India (Garrick et al. 2019).

Urbanization can lead to the deterioration of water quality, with increased point-source and non-point-source pollution from urban surfaces degrading the quality of streams and other receiving waters. In addition, land-use changes can affect the water system in multiple ways (Table 2.4). In some countries, contaminants from industry and untreated wastewater are often pumped directly into waterways due to poor regulatory policies and limited access to sanitation (McGrane 2016; USGS n.d.). Urbanization can lead to sedimentation of waterways when land is cleared for urban development. The exposed soil is prone to erosion and can cause large quantities of sediment to be washed into nearby waterways through surface runoff, leading to decreased water clarity, damage to fish gills, reduced numbers of invertebrates, and increased water treatment costs (McDonald et al. 2016; NIWA n.d.).

Table 2.4 Impacts of land-use change on local water systems.

Stage of Urbanization	Impacts: Change in Land Use	Impacts: Effect on the Water System
Beginning	Removal of trees and vegetation Building of houses, some with sewers and some with septic tanks Drilling of wells	More storm runoff and erosion because there is less vegetation to slow water as it runs down hills More sediment washed into streams Flooding due to changed water drainage patterns
Continuing	Addition of more roads, houses, and commercial and industrial buildings Increase in wastewater discharged into streams Building of new water supply and distribution systems to supply the growing population Construction of reservoirs to provide water Changing of stream channels to accommodate building construction Drilling of deep, large-capacity wells	Less water soaking into the ground due to more impervious surfaces, meaning that the underground water table has less water to recharge it, resulting in a lower water table Drying up of existing wells that are not deep enough to reach the water table Passing of runoff from the increased pavement into storm sewers and on into streams, causing flooding Increased discharge of sewage into streams that do not naturally handle that much water

Urbanization can increase flooding risks, with impermeable surfaces such as roads, roofs, car parks, and footpaths reducing infiltration of the water into the ground and accelerating runoff to ditches and streams, resulting in localized flooding (USGS 2016). For instance, between 1990 and 2015, Edinburgh lost on average 11.27 hectares per annum of vegetated land to urban land cover (from all types of change, including urban creep – the conversion of pervious surfaces to built-up impervious surfaces – and urban expansion). The result is that by 2080, 57 000 additional homes in the city could be at risk of flooding (Rowland et al. 2019).

Rapid Economic Growth

Over the next few decades, population growth and rising income levels will increase global demand for goods and services. Global gross domestic product (GDP) is projected to quadruple between 2011 and 2060. Global materials use is expected to more than double from 79 gigatons (Gt) in 2011 to 167 Gt in 2060, with nonmetallic minerals, such as sand, gravel, and limestone, representing more than half of total materials use (OECD 2019). By 2050, global demand for water for manufacturing, thermal electricity generation, and domestic use will have increased by 400, 140, and 130%, respectively. Demand for water in industry will grow at varying levels globally – for example, by 800% in Africa, where present industry is negligible, compared to 250% in Asia (Boretti and Rosa 2019).

Access to adequate and reliable supplies of water is critical to supporting heavy water-dependent sectors. However, floods, droughts, and other water-related risks can have economic and employment repercussions far beyond the immediately affected areas. In addition to agriculture and industry, sectors with heavily water-dependent jobs include forestry, inland fisheries and aquaculture, mining and resource extraction, water supply and sanitation, and most types of power generation. It is estimated that more than 1.4 billion jobs – or 42% of the world's total workforce – are heavily water-dependent. A further 1.2 billion – or 36% of the workforce – are moderately water-dependent, including jobs in construction, recreation, and transport. Overall, 78% of all jobs in the world are dependent on water (UN Water 2016).

Economic Losses from Floods, Droughts, and Poor Water Quality

Floods and droughts can have direct and indirect effects on a nation's economy, with immediate consequences. Direct effects include damage to infrastructure, while indirect effects include aversion to investing in at-risk areas. Meanwhile, poor water quality can impede economic growth.

Economic Losses from Floods

From an analysis of the impacts of extreme events across 180 countries, spanning the period 1901–2003, evidence shows that the larger the area affected by flooding and drought, the more serious the economic damage: a 1% increase in a region experiencing flooding resulted in a 1.8% decrease in a country's GDP in a single year (Brown et al. 2013). In spring 2019, a significant portion of the United States experienced devastating flooding. The damage was estimated to be $12.5 billion. The main direct impact on GDP came from the effect of flooding on the agricultural sector, with wet weather significantly delaying or reducing the planting of crops such as corn and soybeans; the maturation of these and many other crops was well behind 2018 levels. Indirect losses were attributed to lost work hours and damage sustained by contaminated water (AccuWeather 2019; Federal Reserve Bank of St. Louis 2019).

Economic Losses from Droughts

Likewise, over the period 1901–2003, a 1% increase in a region experiencing drought resulted in a 2.7% decrease in a country's GDP in a single year (Brown et al. 2013). At the micro-level, a study of 78 large Latin American urban areas found that prolonged droughts decreased the level of employment by around 1%, the number of hours worked by around 4.5%, and labor incomes by about 6.5% (Desbureaux and Rodella 2019).

Economic Losses from Poor Water Quality

A World Bank study found that when Biological Oxygen Demand (an index of the degree of organic pollution, and therefore a proxy for overall water pollution) crosses a threshold of 8 milligrams per liter, GDP growth in downstream regions falls significantly – specifically, by 0.83 percentage points, a third of the mean growth rate of the study's location (World Bank 2019).

Water–Energy Nexus

Current water and energy systems are interdependent. Water is used in all phases of energy production and electricity generation, while energy is required to extract, convey, and deliver water of appropriate quality for human uses, and then again to treat wastewater before it is returned to the environment. From an energy perspective, constraints on water can impact the reliability of existing operations, as well as the physical, economic, and environmental viability of future projects. The use of water for energy production can also affect its availability and quality for other users and uses. Meanwhile, the interdependence of water services and the energy sector can impact the ability to provide water- and wastewater-related services (US DOE 2014; IEA 2016).

The energy sector currently accounts for around 15% of global water withdrawals (UN Water 2016). Global energy demand is projected to increase by approximately 27% between 2017 and 2040. At the same time, water withdrawals for the energy sector are expected to rise by less than 2%, to reach over 400 billion cubic meters (bcm), as the amount of water consumed increases by almost 60%, to over 75 bcm, mainly due to advanced cooling technologies in the power sector that withdraw less water but consume more (IEA 2016; UN Water 2019a).

Currently, energy consumption by the global water sector is estimated to be equal to all the energy used by Australia today, most of it in the form of electricity. In 2014, around 4% of global electricity consumption was used to extract, distribute, and treat water and wastewater, along with 50 million tons of oil equivalent of thermal energy, mostly diesel used for irrigation pumps and gas in desalination plants. The International Energy Agency predicts that by 2040, the amount of energy used in the water sector will more than double, with the most substantial increase coming from desalination, followed by large-scale water transfer and increasing demand for wastewater treatment, as well as higher levels of treatment (IEA 2016).

Water–Food Nexus

Over the past 50 years, the world's agricultural production has increased between 2.5 and 3.0 times. More than 40% of the increase in food production has come from irrigated areas. Currently, agriculture accounts for 70% of all water withdrawn from aquifers, streams, and lakes. Since 1961, land under irrigation has more than doubled. Today, the global area equipped for irrigation continues to expand at a rate of 0.6% per annum. Groundwater use in irrigation is increasing, and almost 40% of the irrigated area is now reliant on groundwater either as a primary source or in conjunction with surface water (FAO 2011).

With the world's population expected to reach 9 billion by 2050, demand for food will increase by 60%. The volume of water withdrawn for irrigation globally is expected to grow from 2600 km^3/year in 2005–2007 to 2900 km^3/year in 2030 and almost 3000 km^3/year in 2050. The irrigation requirement, which is the portion of consumptive use from irrigation withdrawals, is estimated to increase from 1270 to 1340 km^3 (FAO 2011, 2015). The impacts of increased irrigation include a reduction in environmental flows, changes in downstream access to water, and a reduction in the extent of wetlands that have essential ecological functions in terms of biodiversity, nutrient retention, and flood control. Groundwater sources are at risk of being mined. In other words, the rate of groundwater extraction is higher than its replenishment rate (Brears 2016b).

Meanwhile, agriculture is one of the most significant contributors to non-point-source pollution impacting rivers, streams, and lakes, as well as wetlands and groundwater supplies. Fertilizer and animal manure, both rich in nitrogen and phosphorous, are the primary sources of nutrient pollution from agricultural sources. Overall, excess nutrients can cause harmful algal blooms, negatively impacting public health, creating dead zones in water, increasing water treatment costs, and affecting industries dependent on clean water (Brears 2016b).

Role of Demand Management and Green Infrastructure in Achieving Water Security

Both demand management and green infrastructure have a crucial role in managing water quantity and water quality to achieve regional water security.

Demand Management

Traditionally, water managers have relied on large-scale, supply-side infrastructural projects such as dams and reservoirs to meet increased demands for water (supply-side management). However, these projects are costly, both economically and environmentally. Also, with most water resources being transboundary, supply-side projects can create political tensions as they rely on water crossing intra- and interstate administrative and political lines. Therefore, there is a need to move toward managing actual demand for water (demand management), as ultimately it is society's attitudes and behaviors toward water that determine the amount that needs supplying. Demand management involves making better use of existing water supplies before attempting to increase them further. Specifically, it promotes water conservation during both normal and abnormal conditions, through changes in practices, culture, and people's attitudes toward water resources. It seeks to reduce the loss and misuse of water, optimize its use, and facilitate significant financial and infrastructural savings by minimizing the need to meet increasing demand with new supplies (Brears 2016a).

Demand management comprises a variety of interventions and organizational systems that intend to improve technical, social, economic, environmental, and institutional efficiencies in the various uses of water. These include:

- Reducing loss and misuse in the various water sectors.
- Optimizing water use by ensuring reasonable allocation between various water users while considering the supply needs of streamflow processes, resource conservation, and quality, as well as the development of *in situ* uses of water, such as recreational use and energy.
- Adding more value per unit of resource mobilized.
- Facilitating significant financial and infrastructural savings for cities and countries, as well as companies.
- Helping reduce the pressure on resources by reducing or halting unsustainable exploitation of both renewable and nonrenewable water sources (GWP 2012).

Green Infrastructure

The most common means of mitigating the risks from climatic extremes has been increasing investment in conventional – or “gray” – infrastructure, such as dams and levees. However, engineers and decision-makers have come to realize the economic and environmental costs of these solutions. For example, economically, gray infrastructure is often capital-intensive in terms of building, operation, maintenance, and replacement. Furthermore, as it is mainly built to address a specific water management problem, it can amplify risks downstream. Environmentally, gray infrastructure often degrades the quality and quantity of water supply, resulting in ecosystem degradation. As such, there has been a turn toward more long-term economically and environmentally sustainable green infrastructure solutions that provide equivalent or similar benefits to gray infrastructure. Green infrastructure involves the use of natural or semi-natural systems that utilize nature’s ecosystem services in the management of water resources and associated risks. Green infrastructure comes in a variety of shapes and sizes and is implemented in a wide variety of contexts, ranging from green and blue roofs at the urban level, which mitigate urban flooding while providing multiple co-benefits (including improving air quality and reducing the urban heat island effect), to the application of natural water retention measures on agricultural land, which in addition to improving water quality, enhances productivity and provides habitat for wildlife. Green infrastructure solutions:

- Increase the water infiltration and storage capacity of wetlands and soils in order to reduce the impacts of floods.
- Can help purify polluted water – the equivalent gray infrastructure solution is wastewater treatment plants.
- Can mitigate droughts by releasing water from natural storage features such as lakes and aquifers for both human and natural use.
- Can appreciate over time with the regeneration of nature and its associated ecosystem services (Brears 2018).

References

Abedin, M.A., A.E. Collins, U. Habiba, and R. Shaw. (2019). “Climate Change, Water Scarcity, and Health Adaptation in Southwestern Coastal Bangladesh.” *International Journal of Disaster Risk Science* 10, no. 1: 28–42.

AccuWeather. (2019). “Flood Damage and Economic Loss in the Heartland to Reach $12.5 Billion This Spring.” https://www.accuweather.com/en/weather-news/flood-damage-and-economic-loss-in-the-heartland-to-reach-12-5-billion-this-spring/333331 (accessed October 9, 2020).

ADB. (2015). “Water-Related Disasters and Disaster Risk Management in the People’s Republic of China.” https://www.adb.org/sites/default/files/publication/177695/water-related-disasters-prc.pdf (accessed October 9, 2020).

ADB. (2016). “Asian Water Development Outlook 2016: Strengthening Water Security in Asia and the Pacific.” https://www.adb.org/sites/default/files/publication/189411/awdo-2016.pdf (accessed October 9, 2020).

Boretti, A. and L. Rosa. (2019). "Reassessing the Projections of the World Water Development Report." *NPJ Clean Water* 2, no. 1: 15.

Brears, R.C. (2016a). *Urban Water Security*. Hoboken, NJ: John Wiley & Sons.

Brears, R.C. (2016b). "The Water–Food Nexus and the Role of Demand Management." *Future of Food: Journal on Food, Agriculture and Society* 4, no. 3: 73–74.

Brears, R.C. (2018). *Blue and Green Cities: The Role of Blue-Green Infrastructure in Managing Urban Water Resources*. London: Palgrave Macmillan.

Brown, C., R. Meeks, Y. Ghile, and K. Hunu. (2013). "Is Water Security Necessary? An Empirical Analysis of the Effects of Climate Hazards on National-Level Economic Growth." *Philosophical Transactions of the Royal Society A: Mathematical, Physical and Engineering Sciences* 371, no. 2002: 20120416.

Chisolm, E.I. and J.C. Matthews. (2012). "Impact of Hurricanes and Flooding on Buried Infrastructure." *Leadership and Management in Engineering* 12, no. 3: 151–156.

Desbureaux, S. and A.-S. Rodella. (2019). "Drought in the City: The Economic Impact of Water Scarcity in Latin American Metropolitan Areas." *World Development* 114: 13–27.

Doocy, S., A. Daniels, S. Murray, and T.D. Kirsch. (2013). "The Human Impact of Floods: A Historical Review of Events 1980–2009 and Systematic Literature Review." *PLoS Currents* 5: ecurrents.dis.f4deb457904936b07c09daa98ee8171a.

EEA. (2018). "Climate Impacts on Water Resources." https://www.eea.europa.eu/archived/archived-content-water-topic/water-resources/climate-impacts-on-water-resources (accessed October 9, 2020).

FAO. (2011). "The State of the World's Land and Water Resources for Food and Agriculture: Managing Systems at Risk." http://www.fao.org/3/i1688e/i1688e.pdf (accessed October 9, 2020).

FAO. (2015). "Towards a Water and Food Secure Future: Critical Perspectives for Policy-Makers." http://www.fao.org/3/a-i4560e.pdf (accessed October 9, 2020).

Federal Reserve Bank of St. Louis. (2019). "How Might Spring Flooding Affect Real GDP?" https://www.stlouisfed.org/on-the-economy/2019/august/how-spring-flooding-affect-real-gdp (accessed October 9, 2020).

Garrick, D., L. De Stefano, W. Yu, I. Jorgensen, E. O'Donnell, L. Turley, et al. (2019). "Rural Water for Thirsty Cities: A Systematic Review of Water Reallocation from Rural to Urban Regions." *Environmental Research Letters* 14, no. 4: 043003.

Gunda, T., D. Hess, G.M. Hornberger, and S. Worland. (2019). "Water Security in Practice: The Quantity–Quality–Society Nexus." *Water Security* 6: 100022.

GWP. (2012). "Water Demand Management: The Mediterranean Experience." http://www.gwp.org/Global/ToolBox/Publications/Technical%20Focus%20Papers/01%20Water%20Demand%20Management%20-%20The%20Mediterranean%20Experience%20(2012)%20English.pdf (accessed October 9, 2020).

Hunter, P.R., A.M. MacDonald, and R.C. Carter. (2010). "Water Supply and Health." *PLOS Medicine* 7, no. 11: e1000361.

IEA. (2016). "WEO-2016 Special Report: Water-Energy Nexus." https://webstore.iea.org/weo-2016-special-report-water-energy-nexus (accessed October 9, 2020).

IPCC. (2014). "The Fifth Assessment Report (AR5) of the Intergovernmental Panel on Climate Change." https://www.ipcc.ch/assessment-report/ar5/ (accessed October 9, 2020).

IWA. (n.d.). "Urbanisation." https://iwa-network.org/urbanisation/ (accessed October 9, 2020).

Kuwayama, Y., A. Thompson, R. Bernknopf, B. Zaitchik, and P. Vail. (2018). "Estimating the Impact of Drought on Agriculture Using the US Drought Monitor." *American Journal of Agricultural Economics* 101, no. 1: 193–210.

McDonald, R.I., K.F. Weber, J. Padowski, T. Boucher, and D. Shemie. (2016). "Estimating Watershed Degradation over the Last Century and Its Impact on Water-Treatment Costs for the World's Large Cities." *Proceedings of the National Academy of Sciences* 113, no. 32: 9117–9122.

McGrane, S.J. (2016). "Impacts of Urbanisation on Hydrological and Water Quality Dynamics, and Urban Water Management: A Review." *Hydrological Sciences Journal* 61, no. 13: 2295–2311.

NCEH. (2016). "Drought and Public Health in the US." https://www.cdc.gov/nceh/hsb/cwh/hsb:drought_infographic508-cleared.pdf (accessed October 9, 2020).

NDMC. (n.d.). "How Does Drought Affect Our Lives?" https://drought.unl.edu/Education/DroughtforKids/DroughtEffects.aspx (accessed October 9, 2020).

NIWA. (n.d.). "Sediment and Urbanisation." https://niwa.co.nz/our-science/freshwater/tools/kaitiaki_tools/land-use/urbanisation/impacts/sedimentation-from-urbanisation (accessed October 9, 2020).

Northland Regional Council. (2011). "Surface Water Quantity." https://www.nrc.govt.nz/resource-library-archive/environmental-monitoring-archive2/state-of-the-environment-report-archive/2011/state-of-the-environment-monitoring/our-freshwater/surface-water-quantity/ (accessed October 9, 2020).

OECD. (2016). "Water Security for Better Lives: A Summary for Policymakers." https://www.oecd.org/env/resources/Water%20Security%20for%20Better%20Lives-%20brochure.pdf (accessed October 9, 2020).

OECD. (2019). "Global Material Resources Outlook to 2060: Economic Drivers and Environmental Consequences." https://www.oecd.org/environment/waste/highlights-global-material-resources-outlook-to-2060.pdf (accessed October 9, 2020).

OECD. (n.d.). "Infographic – What Are the Impacts of Water Pollution?" https://www.oecd.org/water/infographic-impacts-of-water-pollution.htm (accessed October 9, 2020).

Office of the Queensland Chief Scientist. (2018). "What Are the Consequences of Floods?" https://www.chiefscientist.qld.gov.au/publications/understanding-floods/flood-consequences (accessed October 9, 2020).

Rowland, C., P. Scholefield, A. O'Neil, and J. Miller. (2019). "Quantifying Rates of Urban Creep in Scotland: Results for Edinburgh between 1990, 2005 and 2015." https://www.crew.ac.uk/sites/www.crew.ac.uk/files/sites/default/files/publication/Quantifying%20rates%20of%20urban%20creep%20for%20Scotland%20MAIN%20REPORT.pdf (accessed October 9, 2020).

Stanke, C., M. Kerac, C. Prudhomme, J. Medlock, and V. Murray. (2013). "Health Effects of Drought: A Systematic Review of the Evidence." *PLoS Currents* 5: ecurrents.dis.7a2cee9e980f91ad7697b570bcc4b004.

Strosser, P., T. Dworak, P.A.G. Delvaux, M. Berglund, G. Schmidt, J. Mysiak, et al. (2012). "Gap Analysis of the Water Scarcity and Droughts Policy in the EU" https://ec.europa.eu/environment/water/quantity/pdf/WSDGapAnalysis.pdf (accessed October 9, 2020).

UN DESA. (2019a). "World Population Prospects 2019: Highlights." https://population.un.org/wpp/Publications/Files/WPP2019_Highlights.pdf (accessed October 9, 2020).

UN DESA. (2019b). "World Urbanization Prospects: The 2018 Revision." https://population.un.org/wup/Publications/Files/WUP2018-Report.pdf (accessed October 9, 2020).
UNISDR. (2015). "The Human Cost of Weather-Related Disasters 1995–2015." https://www.unisdr.org/we/inform/publications/46796 (accessed October 9, 2020).
UN Water. (2013). "Water Security and the Global Water Agenda." https://www.unwater.org/publications/water-security-global-water-agenda/ (accessed October 9, 2020).
UN Water. (2016). "Water and Jobs." https://unesdoc.unesco.org/ark:/48223/pf0000244040/PDF/244040eng.pdf.multi (accessed October 9, 2020).
UN Water. (2019a). "UN-Water Policy Brief on Climate Change and Water." https://www.unwater.org/publications/un-water-policy-brief-on-climate-change-and-water/ (accessed October 9, 2020).
UN Water. (2019b). "World Water Development Report 2019: Leaving No One Behind." https://knowledge.unccd.int/publications/world-water-development-report-2019-leaving-no-one-behind (accessed October 9, 2020).
UN Water. (n.d.). "Water and Climate Change." https://www.unwater.org/water-facts/climate-change/ (accessed October 9, 2020).
US DOE. (2014). "The Water–Energy Nexus: Challenges and Opportunities." https://www.energy.gov/sites/prod/files/2014/07/f17/Water%20Energy%20Nexus%20Executive%20Summary%20July%202014.pdf (accessed October 9, 2020).
US EPA. (2017). "Climate Impacts on Water Resources." https://19january2017snapshot.epa.gov/climate-impacts/climate-impacts-water-resources_.html (accessed October 9, 2020).
USGS. (2016). "Effects of Urban Development on Floods." Fact Sheet 076-03. https://pubs.usgs.gov/fs/fs07603/ (accessed October 9, 2020).
USGS. (n.d.). "Urbanization and Water Quality." https://www.usgs.gov/special-topic/water-science-school/science/urbanization-and-water-quality?qt-science_center_objects=0#qt-science_center_objects (accessed October 9, 2020).
World Bank. (2019). "Quality Unknown: The Invisible Water Crisis." https://openknowledge.worldbank.org/handle/10986/32245 (accessed October 9, 2020).
World Resources Institute. (2020). "The Number of People Affected by Floods Will Double Between 2010 and 2030." https://www.wri.org/blog/2020/04/aqueduct-floods-investment-green-gray-infrastructure (accessed October 9, 2020).

Part I

Demand Management

3

Water Allocation

Introduction

Growing water scarcity, climatic extremes, and increasing water demand for economic development have led to a rising global interest in water allocation practices. In particular, an increasing number of locations around the world are reforming their water allocation regimes to manage the risk of shortage and to adjudicate between competing users via a combination of policies, laws, and mechanisms (OECD 2015; UNECE 2017). This chapter will first provide an overview of the objectives of water allocation planning, before turning to look at various allocation tools, including water pricing, subsidies, and water trading mechanisms. Finally, it will discuss the role of basin-level modeling, seasonal forecasting, and water restrictions in allocating water resources.

Objectives of Water Allocation Planning

Water allocation is the process whereby an available water resource is distributed (or redistributed) to legitimate claimants. Water allocation planning is typically undertaken to achieve a series of key policy objectives, which include:

- *Equity*: Allocating water in a way that is fair and equitable among different regions and user groups. This can consist of equity between different administrative regions or between upstream and downstream areas.
- *Environmental Protection*: Allocating water in a way that recognizes the needs of freshwater-dependent ecosystems and protects essential freshwater services such as groundwater recharge.
- *Development Priorities*: Allocating water in a way that promotes economic and social development.
- *Balancing Supply and Demand*: Allocating water in a way that balances water supplies with water demands, particularly in order to manage the natural variability of water availability and to avoid frequent or unexpected water deficits.
- *Efficient Use of Water*: Allocating water in a way that supports its most efficient use (ADB 2008, 2013).

Regional Water Security, First Edition. Robert C. Brears.

Water Pricing

Pricing of water and water-related services is an essential mechanism for managing demand, promoting the efficient use of water, and allocating water among competing uses. The following pricing structures encourage water conservation and water efficiency:

- *Increasing Block Rate*: Using block rates or tiered pricing that increases with water usage. The per-unit charge for water increases as the amount of water used increases. The first block is charged at one rate, the next block is charged at a higher rate.
- *Time of Day Pricing*: Charging higher prices for water used during a utility's peak demand periods.
- *Water Surcharges*: Charging a higher rate for "excessive" water use, i.e. water consumption that exceeds the local or regional average.
- *Seasonal Rates*: Water prices rise or fall according to weather conditions and the corresponding demand for water.

These pricing structures are less effective in encouraging water conservation, however:

- *Uniform Rate Structures*: A uniform rate charges the same price per unit of water usage beyond the fixed customer charge, which covers some fixed costs. The water bill will vary with usage. Uniform rates by class charge the same price per unit for all customers within a customer class (e.g., residential or nonresidential).
- *Flat Fee Rates*: Flat fee rates do not vary by customer characteristics or water usage (Beecher and Kalmbach 2013; Ben Zaied et al. 2017; Ratnasiri et al. 2018; Soto Rios et al. 2018; US EPA n.d.).

For conservation pricing structures to be effective, customers should be educated on the type of rate structure the utility uses and be provided regular feedback on their usage through their water bill, which should be sent monthly. Most customers do not track their water use between bills because of the difficulty and inconvenience of reading a meter. When they read their bill, they usually only take note of the total amount. Therefore, conservation pricing provides more robust feedback to customers, who will often see a more considerable percentage increase in their water bill than in water use. Water utilities should adopt billing software that allows customers to compare the water use on their bill with the average use for their customer class, as well as view their individual water use for the past 12 months. The rate structure should be clearly indicated (Texas Water Development Board 2004; Castillo-Martinez et al. 2014).

Cost Recovery

The pricing of water can be used as a financial tool to recover the direct (water supply and infrastructure) and indirect (environmental, social, and opportunity) costs of providing water and water-related services. There are two main types of cost recovery:

- *Supply Cost Recovery*: Recovery of the financial costs of water supply, including investments in infrastructure, operational and maintenance costs, and administrative costs.

- *Full Cost Recovery*: Recovery of financial as well as water use-related environmental, social, and opportunity costs. If environmental, social, and opportunity costs are not integrated (internalized) into prices, they are referred to as "external costs" or "externalities."

The meaning and level of cost recovery depend on what is considered part of the "price" of providing and using water. Often, external costs are more difficult to define and quantify, so they are rarely considered in water tariffs (European Commission 2012). Nonetheless, when consumers do not pay the full cost of water, they often use it inefficiently. At the same time, when price levels are too high, this may pose problems of continued access and affordability for poorer consumers (Savenije and Van Der Zaag 2002; OECD 2013).

Traditionally, water is viewed as cheap and plentiful, so water prices are often set to achieve only partial cost recovery, with a rate of return frequently below commercial levels. This "underpricing" of water is commonly justified based on egalitarianism, where water is seen as a natural asset indispensable for human survival, and so as one which should be priced low for low-income consumers to ensure social equity. However, water is no longer a natural asset. Instead, its current use requires transport, treatment, and distribution, resulting in substantial (environmental and economic) costs of provision. Nevertheless, these costs are seldom fully factored into the price of water, leading to frequent underpricing, which promotes overconsumption, intensifying water scarcity and thus further raising the costs of using water resources. This chain of events adversely impacts the sustainable use of water in two ways. First, the actual total amount of water used exceeds sustainable levels, creating welfare losses for current and future generations. Second, the increasing costs induced by an increase in scarcity lead to further welfare losses as water used for numerous economic and social activities needs to be restricted.

As a result of egalitarian underpricing of water, low-income households will thus be more intensely affected in the future by rising water prices due to increased scarcity. Because any price increase will be greater than today's full economic cost of providing water, therefore, future prices will take a greater percentage of household income from poorer households compared with wealthier ones. As such, water prices today need to reflect the full costs of providing water, including environmental, social, and supply costs, to ensure efficient and sustainable use prevails and so avoid future price increases. Nonetheless, the price of water should be based on a "reasonable price structure" that not only aims at cost recovery but also considers access to safe water for the poor and the state of the natural environment from which the water is drawn. Setting a reasonable price sends a clear signal to end users that water should be used wisely (Brears 2016).

To promote the sustainable use of water while ensuring affordability and preventing water poverty, two different types of policies are available: income-support policies and tariff-related policies. Income-support programs involve the provision of subsidies directly to the most vulnerable households. Tariff-related policies, on the other hand, promote cross-subsidization between poor households and wealthy ones, with large-volume water users paying more than those who use only small (baseline) amounts (Molinos-Senante and Donoso 2016).

Raising Revenue

In addition to encouraging water conservation, conservation-based pricing structures can be used to raise revenue for water utilities facing economic and population growth, aging infrastructure, and increasingly stringent environmental regulations. To meet these needs, water and wastewater providers must increase their investments, with rising capital needs attributed to:

- Sewage treatment plants and underground pipe systems needing replacement as their useful lives expire.
- More stringent drinking water and wastewater standards.
- Increasing expenses associated with developing new sources of water (US EPA 2011; Inglezakis et al. 2016).

Price Increases

When water prices need to be raised, or the pricing structure modified, acceptance is improved when the public understands:

- The need for capital improvement programs such as expansions and upgrades.
- The benefits associated with water conservation (e.g., pollution prevention through reduced water withdrawals and wastewater flows, habitat protection, energy conservation).
- The need for conservation measures such as metering, improved water accounting, leak detection, water-use audits, retrofits, reuse and recycling, and landscape improvements.

Overall, the strategic pricing of water and wastewater can slow or manage demand and raise revenue for capital investments (US EPA 2011).

Case 3.1 Singapore's Water Price Revision

Singapore's Public Utilities Board (PUB) is continuing to invest in and upgrade its water supply system to make it more resilient, with an emerging concern being the effect of climate change on the water supply. At the same time, the costs of water treatment, reservoir operations, NEWater (ultra-clean, high-grade reclaimed water) production, and maintenance and expansion of the island-wide network of water pipelines are increasing. In 2000, the water system cost around $500 million to operate. By 2015, this had risen to $1.3 billion. Over the period 2017–2021, PUB expects to invest another $4 billion in water infrastructure, including investments in NEWater and desalination factories, pipelines and pumping stations, and water reclamation plants. To meet the higher costs of treating and supplying water and ensure that prices reflect the scarcity value of water, a price revision was undertaken over two phases on July 1, 2017 and July 1, 2018. The revised prices for domestic, nondomestic, NEWater, and industrial water users are detailed in Tables 3.1, 3.2, 3.3, and 3.4, respectively (Government of Singapore 2017; PUB n.d.).

Table 3.1 Potable water price for domestic customers (households).

	Phase 1: From July 1, 2017		Phase 2: From July 1, 2018	
	Price ($/m³)		Price ($/m³)	
Monthly Water Usage	*0–40 m³*	*>40 m³*	*0–40 m³*	*>40 m³*
Tariff	$1.19	$1.46	$1.21	$1.52
Water conservation tax (% of tariff)	$0.42 (35% of $1.19)	$0.73 (50% of $1.46)	$0.61 (50% of $1.21)	$0.99 (65% of $1.52)
Waterborne fee	$0.78	$1.02	$0.92	$1.18
Total price	**$2.39**	**$3.21**	**$2.74**	**$3.69**

Source: Government of Singapore, "Why Increase Water Prices?," https://www.gov.sg/article/why-increase-water-prices; PUB, "Water Price," https://www.pub.gov.sg/watersupply/waterprice

Table 3.2 Potable water price for non-domestic customers (businesses).

	Phase 1: From July 1, 2017	Phase 2: From July 1, 2018
	Price ($/m³)	Price ($/m³)
Tariff	$1.19	$1.21
Water conservation tax (% of tariff)	$0.42 (35% of $1.19)	$0.61 (50% of $1.21)
Waterborne fee	$0.78	$0.92
Total price	**$2.39**	**$2.74**

Source: Government of Singapore, "Why Increase Water Prices?," https://www.gov.sg/article/why-increase-water-prices; PUB, "Water Price," https://www.pub.gov.sg/watersupply/waterprice.

Table 3.3 NEWater price.

	Phase 1: From July 1, 2017	Phase 2: From July 1, 2018
	Price ($/m³)	Price ($/m³)
Tariff	$1.28	$1.28
Water conservation tax (% of tariff)	$0.13 (10% of $1.28)	$0.13 (10% of $1.28)
Waterborne fee	$0.78	$0.92
Total price	**$2.19**	**$2.33**

Source: Government of Singapore, "Why Increase Water Prices?," https://www.gov.sg/article/why-increase-water-prices; PUB, "Water Price," https://www.pub.gov.sg/watersupply/waterprice.

Table 3.4 Industrial water price.

	Phase 1: From July 1, 2017	Phase 2: From July 1, 2018
	Price ($/m³)	Price ($/m³)
Tariff	$0.66	$0.66
Waterborne fee	$0.78	$0.92
Total price	**$1.44**	**$1.58**

Source: Government of Singapore, "Why Increase Water Prices?," https://www.gov.sg/article/why-increase-water-prices; PUB, "Water Price," https://www.pub.gov.sg/watersupply/waterprice.

Subsidies

The World Bank defines subsidies as a subset of funding flows between governments, service providers, and customers. They occur when a user/customer pays less for a product or service than the service provider's cost, leaving a third party, such as a government, responsible for converting the difference. Subsidies come in many forms, including financial transfers between two entities (e.g., a utility and a customer) and implicit transfers, which occur when products, services, or inputs are underpriced (World Bank 2019). Subsidies are widely used by governments to encourage more environmentally beneficial behavior (e.g., the providing of a rebate on the purchase of a water-efficient technology) (INTOSAI Working Group on Environmental Auditing 2016). Nonetheless, around the world, water-related subsidies commonly fail to achieve their objectives due to poor design; they tend to be overly pervasive, expensive, poorly targeted, nontransparent, and distortionary, and to serve only political interests. The following principles should be followed to ensure subsidies meet their objectives:

- *Subsidies Should Achieve the Intended Policy Outcome*: Subsidies require a smart design and clarity about their policy objectives and short- and long-term goals.
- *Subsidies Should Reach the Intended Target Groups*: They require clarity on who is the intended target group and how they can best be reached. They also require rigorous monitoring to track how they are reaching them.
- *Subsidies Should Be Financially Sustainable*: Subsidies require a thorough understanding about the potential costs of the program. This includes both upfront capital costs and long-term operational and maintenance costs.
- *Subsidies Should Integrate Local Peoples' Needs*: To guarantee the sustainability of the subsidized water technology, it is of prime importance to facilitate the integration and participation of the local beneficiaries and to develop a sense of ownership toward the new infrastructure.
- *Subsidies Should Be Implemented Clearly and Transparently*: As subsidies involve public funds, subsidy programs need to be clear and transparent, enabling eligible households or communities to access them and providing clear recourse mechanisms in cases where there is a suggestion of impropriety (WSSCC 2009).

Implementing Subsidies

To implement a successful subsidy scheme, there needs to be a legal framework at its base, an institutional framework charged with its implementation, and a series of administrative procedures to govern the operation of the system. Historically, subsidies have suffered from a lack of clarity regarding their objectives and lack of transparency regarding their procedures. The result is that they have been provided with a high degree of discretion, mainly to further political ends. Therefore, subsidy systems should be given a clear basis in law, which would serve to strengthen the legitimacy of the subsidy program and to ensure that funds were used in a socially useful manner. The law should establish the following points:

- The government institution with overall responsibility for the subsidy program.
- The potential sources of finance for the subsidy program.
- The basic principles that should be observed in administering subsidies.

The chief administrative procedures required to operationalize a subsidy scheme are summarized in Table 3.5 (SSWM n.d.). Overall, if well designed, subsidies can be powerful and progressive tools that ensure all people benefit from water supply and sanitation services. Some suggestions are listed in Table 3.6 (World Bank 2019).

Table 3.5 Administrative procedures to operationalize subsidy schemes.

Administrative Procedure	Description
Budgeting	The scheme should incorporate a system for projecting likely costs from year to year and should identify how any shortfall or surplus of funds should be dealt with
Selection	The scheme should establish the routes through which applications can be made, how they will be screened, and the process for reaching a final decision regarding eligibility for a subsidy
Payment	The scheme should clarify how subsidy beneficiaries will be billed, and establish how and when funds will be transferred from the government to the concessionaire

Source: Modified from SSWM, "Subsidies (Wd)", https://sswm.info/water-nutrient-cycle/water-distribution/softwares/economic-tools/subsidies-%28wd%29.

Table 3.6 Recommendations to ensure subsidies are effective.

Smart	Targeted	Implemented Effectively
Most subsidies go to water, urban, and networked services. A better balance across water and sanitation, rural and urban, and different types of service can make subsidies work harder Subsidies can encourage better operational efficiency through performance incentives A single instrument is unlikely to attain all policy goals simultaneously	Measures to make water supply and sanitation affordable for those in need can ensure that no one gets left behind Effective targeting is increasingly possible through technological innovation	A communications strategy is essential to build advance backing and for successful implementation Understanding the institutions, incentives, and interests that shape subsidy reform is vital to cultivating supportive political coalitions When a subsidy is temporary, an appropriate exit strategy must include some form of support for the most vulnerable Complementary policy measures can make scarce public resources go further

Source: Modified from World Bank, "Doing More with Less - Smarter Subsidies for Water Supply and Sanitation", https://www.worldbank.org/en/topic/water/publication/smarter-subsidies-for-water-supply-and-sanitation

Case 3.2 Toronto's Water Efficiency for Business Rebate

Toronto Water provides rebates to help businesses reduce their water use. By lowering water consumption, companies can lower operating costs (less wastewater, chemicals, and energy costs), enhance their public image by demonstrating a commitment to the environment, and increase their competitiveness by saving money and attracting new customers. Several rebates are offered to businesses:

- *Capacity Buyback Program*: This program provides a free water audit to help customers identify ways of reducing water use and rewards participants that implement permanent water-saving measures with a one-time cash rebate of up to 30 cents per liter of water saved per average day. It helps participants save money over the long term through reduced water bills.
- *Industrial Water Rebate Program*: This program (also called the Block 2 water rate) offers a discounted water rate to manufacturers in Toronto in order to support economic growth and encourage water conservation. It is open to manufacturers that use more than 5000 cubic meters of water annually, fall within the industrial property tax class, are in full compliance with Toronto's Sewer By-law, and submit a comprehensive plan to the satisfaction of Toronto Water. Customers are charged the general water rate on the first 5000 cubic meters of water use; any use over that amount is eligible for a rate reduction of 30%. In 2019, the difference in the water rate was:
  - *General Water Rate (Block 1 Rate)*: $3.9549 per cubic meter.
  - *Industrial Water Rate (Block 2 Rate)*: $2.7684 per cubic meter.
- *Sewer Surcharge Rebate Program*: Businesses in Toronto may be eligible for a rebate on water not discharged into the sanitary sewer system (i.e., water evaporating from cooling towers or used to make a product). The rebate is based on the percentage of water not returned to the sanitary sewer system and is applied to the sewer portion of the water rate (Table 3.7) (Toronto Water n.d.).

Table 3.7 Toronto Water's sewer surcharge rebate program.

Property's Total Annual Water Consumption (Cubic Meters)	Sanitary Contribution Amount
≤1500	≤80%
>1500 but ≤15 000	≤85%
>15 000 but ≤1 500 000	≤90%
>1 500 000	≤95%

Source: Based on Toronto Water, "Water Efficiency for Business", https://www.toronto.ca/services-payments/water-environment/how-to-use-less-water/water-efficiency-for-business/

Water Trading (Reallocation)

Water trading can be defined as the voluntary buying and selling of water in some quantifiable form, either now or in the future. Water markets are built on the "currency" of water rights (i.e., the right to use a specific volume of water each year). Table 3.8 lists the typical

Table 3.8 Typical attributes of a water right.

Attribute	Description
Quantity	The amount of water the holder of the water right may withdraw and/or consumptively use, or the area of land and crops that can be irrigated
Source	The specific water source and location for which the water right is granted
Timing	Restrictions on the times when the water right applies (i.e., times that the volume may be withdrawn or consumptively used, and possibly daily extraction limits)
Assurance	How absolute the water right is: some are more absolute (volume is almost always fulfilled), while others have a variable assurance of supply depending on how much water is available each year
Type of use	The specific use for which the water is to be withdrawn or consumptively used (e.g., irrigation, mining, etc.)
Duration	How long the holder is entitled to the water right: some rights are issued in perpetuity, while others are authorized only for a specific length of time (e.g., months or years)

attributes of water rights. These are usually issued by an agency of government, such as a state-, provincial-, or national-level water management agency. A water right is typically an entitlement to use water, but it does not convey ownership of the water itself. Therefore, water remains a public resource, owned and managed by governments in the public interest. This enables governments the opportunity to adjust or restrict the volume of water that can be used during times of shortage.

The terminology of the water currency being exchanged in a water market differs across geopolitical boundaries and even within countries. For example, in the Western United States, a permanent water right can be sold or temporarily leased. In Australia, permanent rights are referred to as "water entitlements," and entitlement holders receive differing volumes of "water allocations" each year according to water availability (The Nature Conservancy 2016).

A water market brings together willing buyers and sellers that want to exchange water rights, with buyers seeking the right to use more water and sellers willing to trade some of their rights for monetary compensation. Using Australian terminology, three types of water trading may occur in a market:

1) Short-term or temporary transfers of water that is already allocated and available for immediate use.
2) Medium-term leasing of water allocations in a manner that enables a water user to plan secure access to water for some time.
3) Permanent transfers of water entitlements (the ongoing property right to either a proportion or a fixed quantity of the available water at a given source) (NCCARF 2013; Wheeler et al. 2017).

Economic Efficiency

The economic advantage of water markets is that they should enable limited resources to be put to their most productive uses (i.e., economic efficiency) by distributing them to those that value them most highly over the short and long term. Using Australian terminology again, this reallocation results in three types of economic efficiency, dependent on the timeframe involved:

1) *Allocative Efficiency*: Changes in water resource demand or use motivated by seasonal conditions, commodity price adjustments, cropping choices, and other short-term decision-making requirements are most often achieved through water allocation trade.
2) *Dynamic Efficiency*: Changes in water resource demand or use that stem from structural alterations such as new investment opportunities, regulatory shifts in access arrangement (e.g., extraction limits), or personal strategic choices are most often achieved through water entitlement trade.
3) *Productive Efficiency*: Changes in the price of both water entitlements and allocation offer incentives for the efficient use of water resources as either an investment or an input for productive outcomes.

As such, water markets require four key elements to drive efficient use and outcomes:

1) *Set Consumptive Pool*: There should be a fixed limit to resource availability – one that is ideally credible and based on accurate science, monitored and enforced, and consistent with sustainable levels of extraction.
2) *Entitlement Shares*: Once the consumptive pool is set, users should be provided with entitlement shares in the form of secure property rights to that pool. Importantly, these shares cannot exceed the total limit.
3) *Trading*: The entitlement shares and the water allocated to them each season should be tradeable under low transaction costs and entry/exit barrier conditions such that ownership, control, and use can change over time.
4) *Price*: Prices for these shares and allocations that consider externality costs to third parties should be established in the market using the value placed on water use by well-informed buyers and sellers (NCCARF 2013).

Overall, there are six main benefits of water trading:

1) *Stimulation of Water Savings*: By establishing a monetary value for water, water markets can provide a strong stimulus for reducing consumptive water use, because a water-saving entity can be rewarded financially by selling or leasing the portion of its water rights that it no longer needs. When water is appropriately priced, this discourages waste.
2) *Increase in Water Availability*: By providing access to additional water through the market, trading allows a community or government to avoid expensive, time-consuming, and environmentally damaging methods of increasing its water supplies.
3) *Improvement of Community Flexibility*: By providing the opportunity to transfer water between users, trading allows individuals and communities to adapt quickly to changing conditions, personal preferences, and needs.
4) *Improvement of Water's Productivity and Allocation Efficiency*: By discouraging wasteful or low-value uses of water, trading facilitates reallocation of water rights to more productive uses, commonly resulting in more revenue generation in local economies.

5) *Return of Water to Nature*: By providing the opportunity to purchase water, markets offer conservation interests and government agencies the ability to restore water flows in depleted freshwater and estuarine ecosystems.
6) *Improvement in Accounting for Water Use and Availability*: When water is appropriately priced and water assets are being traded, water users will be more willing to participate in transparent water measurement and reporting practices (The Nature Conservancy 2016).

Case 3.3 Water Markets and Trading in the Murray–Darling Basin, Australia

Water in the Murray–Darling Basin can be bought or sold, either permanently or temporarily. It may be traded either within catchments, between catchments (when possible), or along river systems. Trading helps farmers in the Basin use water more productively and contributes to sustainable water management.

- Irrigated agriculture in the Basin consumes about 60% of Australia's available water.
- During the year, water is distributed (allocated) against entitlements in response to factors including rainfall and storage levels (there are more than 150 classes of water entitlements in the Basin).
- The water markets create incentives for water to be moved to higher-value uses.
- When water is allocated to an entitlement holder, it can be used as needed (it is the entitlement holder's business decision to make).
- Each year, water trading in the Basin is worth around $2 billion.

The water markets enable a range of water entitlements and water allocations to be bought and sold, where:

- Water entitlements are rights to an ongoing share of water within a system, with the financial value determined by the market.
- Water allocations are the amount of water distributed to users (water entitlement holders) in a given year, with allocations against entitlements changing according to rainfall, inflows into storages, and how much water is already stored.

For buyers and sellers, water trading can increase productivity, allow choice and flexibility in business decision-making, and reduce risk. The New South Wales, Queensland, South Australian, and Victorian governments are primarily responsible for managing water markets, and each state has its own process and rules for allocating water. There are two types of water trade available on these markets:

- *Permanent Trade*: The trade of water entitlements. For example, an entitlement holder might sell its entitlement.
- *Temporary Trade*: The trade of water allocations. For example, an entitlement holder might sell its allocation in a given season, based on its business model (Murray–Darling Basin Authority 2020).

Source: Murray Darling Basin Authority, "Water Markets and Trade", Retrieved from https://www.mdba.gov.au/managing-water/water-markets-and-trade

Basin-Level Modeling and Seasonal Forecasting

With increasing competition for water across regions and sectors, the river basin is recognized as the appropriate unit of analysis for addressing water resource management challenges, where a river basin system is made up of water source components, instream and off-stream demand components, and intermediate (treatment and recycling) components (Cai et al. 2006). Basin-level modeling for water allocation helps optimize water allocation among competing water uses and users in response to changing water availability (seasonal or long-term). The objectives of modeling exercises for water allocation are to determine how best to optimize available resources among competing uses and assess potential trade-offs while ensuring there is enough water left for the environment. By modeling various allocation scenarios, water managers can allocate scarce resources in ways that are both environmentally and economically beneficial. Examples of model predictions include:

- Available seasonal water supply.
- Sustainable extraction rates, to ensure environmental flows.
- Minimum amount of water required for irrigation.
- Minimum requirements for hydropower generation needs.
- Economic benefits of the allocation of water for specific priority uses.
- Implications of future climate change on the availability of water resources, to help inform infrastructure development and land use planning decisions.

According to the UNEP-DHI Partnership, modeling can involve a combination of climatic, hydrologic, and economic data. For example, modeled seasonal water availability, based on past trends and future climate predictions, can be coupled with different water use scenarios to assess the environmental and economic impacts of varying allocation decisions in a basin. The datasets required to inform analysis include hydrological data, such as existing water use trends, the requirements of various users, environmental water requirements, and socioeconomic data, such as population growth trends and planned water infrastructure. Scenarios can also be informed by sociotechnical variables such as improved water efficiency, the water rights of certain users, and expected changes in water quality, depending on resource and data availability. Overall, basin-level modeling and seasonal forecasting provide a range of benefits, as listed in Table 3.9 (CTCN 2017).

Table 3.9 Basin-level modeling and seasonal forecasting benefits.

Benefit Type	Description
Environmental	Helps ensure that a sufficient amount of water is allocated for the environment (ensuring environmental flows), minimizing the risk of resource over-abstraction, pollution, and ecosystem degradation
Socioeconomic	Helps avert conflict resulting from uncontrolled competing uses of scarce resources
	Supports equal distribution among various users, ensuring minimum water requirements are met for human use, the economy, and the environment
	Modelling exercises can help determine the most efficient applications of water that create the most benefit to society (economic or other)
	Adaptive allocation improves climate resilience and mitigates risks of water shortages resulting from changing resource availability and uncoordinated use

Source: Modified from CTCN, "Basin Level Modelling and Seasonal Forecasting for Water Allocation," (2017).

Case 3.4 New Zealand's Water Model

New Zealand's National Institute of Water and Atmospheric Research (NIWA) is leading a partnership with GNS Science, Manaaki Whenua – Landcare Research, and the regional councils of Southland, Horizons, and Gisborne to develop a computer modeling framework that will enable users to accurately predict how much freshwater is available, where it comes from, and how quickly it moves through catchments. The New Zealand Water Model – Hydrology (NZWaM-Hydro), being developed over the period 2016–2022, will enable the prediction of hydrological flow and transport processes from the national to the subcatchment scale. It will provide the ability to predict how climate, land use, and other environmental changes will affect aquatic ecosystems and the availability of water resources for use by businesses and communities. It will also enable communities to develop sustainable use of the resource and better prepare for potential floods and droughts (NIWA 2018).

Source: NIWA, "Nz Water Model - Hydrology," 12 Nov 2018. Retrieved from https://niwa.co.nz/freshwater-and-estuaries/research-projects/nz-water-model-hydrology-nzwam-hydro

Water Restrictions

There are two types of water restriction available to mitigate water source depletion and degradation and ensure environmental water needs are met: temporary restrictions and permanent restrictions. Temporary restrictions are enacted to restrict certain types of water use during specified times or to restrict the total level of water use to a specified amount. These programs are usually adopted during times of severe water shortage and cease once the shortage has passed. Examples include restrictions on nonessential water uses (watering lawns, washing cars, filling swimming pools, and washing driveways), restrictions on commercial use (car washes, hotels, and other large consumers of water), bans on using water of drinking quality for cooling purposes, and bans on non-recirculating carwashes, laundries, and fountains. Permanent water restrictions include amendments to building codes or ordinances requiring the installation of water-saving devices (e.g., low-flow toilets, showerheads, and faucets) in all newly constructed or renovated homes and offices. For instance, plumbing codes can be used to ensure new homes and offices have maximum water-use standards for plumbing fixtures such as toilets, urinals, faucets, and showers (Brears 2016).

Case 3.5 The City of Pasadena's Watering Schedule

The City of Pasadena, California, has prohibited water waste in the following instances:

- Outdoor watering is limited to Tuesday, Thursday, and Saturday in the spring/summer (April 1–October 21), and is restricted to one day a week on Saturday in the fall/winter (November 1–March 31) and must be done before 9 a.m. and after 6 p.m.
- All property owners must fix leaks, breaks, or other sprinkler system malfunctions when they find them, or within 72 hours of receiving a notice from Pasadena Water and Power.

Pasadena also has Permanent Water Waste Prohibitions in place, which include:

- No watering outdoors between 9 a.m. and 6 p.m., except with a hand-held container or hose with a shut-off nozzle, or for short periods when adjusting a sprinkler system.
- No watering during and within 48 hours after periods of significant rain.
- No water flow or runoff. Watering or irrigating any lawn, landscape, or other vegetated areas in a manner that causes or allows runoff such that water flows on to adjacent property, non-irrigated areas, private and public walkways, roadways, parking lots, or structures is prohibited.
- No excessive water flow or runoff on to the pavement, gutters, or ditches from watering or irrigating landscapes or vegetation of any kind.

Pasadena Water and Power will work closely with customers to provide information and issue written warnings for first offences. However, customers who continue to violate the ordinance will face fines for future violations. Fine amounts range from $100 to $500 and are double for nonresidential customers with meters of 1½″ or larger (City of Pasadena 2019).

Source: City of Pasadena, "Pasadena Watering Schedule," Retrieved from: https://ww5.cityofpasadena.net/water-and-power/wateringschedule/. © 2019 Pasadena Water and Power

References

ADB. (2008). "Water Rights and Water Allocation: Issues and Challenges for the Region." https://www.gwp.org/globalassets/global/toolbox/references/water-rights-and-water-allocation.-issues-and-challenges-for-the-asian-region-adb-2008.pdf (accessed October 9, 2020).

ADB. (2013). "Basin Water Allocation Planning: Principles, Procedures and Approaches for Basin Allocation Planning." https://www.adb.org/publications/basin-water-allocation-planning (accessed October 9, 2020).

Beecher, J.A. and J.A. Kalmbach. (2013). "Structure, Regulation, and Pricing of Water in the United States: A Study of the Great Lakes Region." *Utilities Policy* 24: 32–47.

Ben Zaied, Y., H. Bouzgarrou, N. Ben Cheikh, and P. Nguyen. (2017). "The Nonlinear Progressive Water Pricing Policy in Tunisia: Equity and Efficiency." *Environmental Economics and Policy Studies* 8, no. 2: 17–27.

Brears, R.C. (2016). *Urban Water Security*. Hoboken, NJ: John Wiley & Sons.

Cai, X., C. Ringler, and M.W. Rosegrant. (2006). *Modeling Water Resources Management at the Basin Level: Methodology and Application to the Maipo River Basin*. Washington, DC: International Food Policy Research Institute.

Castillo-Martinez, A., A. Gutierrez-Escolar, J.-M. Gutierrez-Martinez, M.J. Gomez-Pulido, and E. Garcia-Lopez. (2014). "Water Label to Improve Water Billing in Spanish Households." *Water* 6, no. 5.

City of Pasadena. (2019). "Pasadena Watering Schedule." https://ww5.cityofpasadena.net/water-and-power/wateringschedule/ (accessed October 9, 2020).

CTCN. (2017). "Basin Level Modelling and Seasonal Forecasting for Water Allocation." https://www.ctc-n.org/sites/www.ctc-n.org/files/resources/basin_level_modelling_seasonal_forecasting_0.pdf (accessed October 9, 2020).

European Commission. (2012). "The Role of Water Pricing and Water Allocation in Agriculture in Delivering Sustainable Water Use in Europe – Final Report." https://ec.europa.eu/environment/water/quantity/pdf/agriculture_report.pdf (accessed October 9, 2020).

Government of Singapore. (2017). "Why Increase Water Prices?" https://www.gov.sg/article/why-increase-water-prices (accessed October 9, 2020).

Inglezakis, V.J., S.G. Poulopoulos, E. Arkhangelsky, A.A. Zorpas, and A.N. Menegaki. (2016). "Aquatic Environment." In *Environment and Development*, edited by S.G. Poulopoulos and V.J. Inglezakis. Amsterdam: Elsevier, pp. 137–212.

INTOSAI Working Group on Environmental Auditing. (2016). "Market Based Instruments for Environmental Protection and Management." https://www.eurosaiwgea.org/documents/meetings/14%20AM/14AM_report_0602.pdf (accessed October 9, 2020).

Molinos-Senante, M. and G. Donoso. (2016). "Water Scarcity and Affordability in Urban Water Pricing: A Case Study of Chile." *Utilities Policy* 43: 107–116.

Murray–Darling Basin Authority. (2020). "Water Markets and Trade." https://www.mdba.gov.au/managing-water/water-markets-and-trade (accessed October 9, 2020).

Nature Conservancy, The. (2016). "Investing Our Way out of the Global Water Crisis." https://www.nature.org/en-us/what-we-do/our-insights/perspectives/investing-our-way-out-of-the-global-water-crisis/ (accessed October 9, 2020).

NCCARF. (2013). "The Role of Water Markets in Climate Change Adaptation." https://www.nccarf.edu.au/sites/default/files/attached_files_publications/Loch_2013_Role_of_water_markets.pdf (accessed October 9, 2020).

NIWA. (2018). "NZ Water Model – Hydrology." https://niwa.co.nz/freshwater-and-estuaries/research-projects/nz-water-model-hydrology-nzwam-hydro (accessed October 9, 2020).

OECD. (2013). "Water Pricing for Public Supply." https://www.oecd-ilibrary.org/environment/environment-at-a-glance-2013/water-pricing-for-public-supply_9789264185715-10-en (accessed October 9, 2020).

OECD. (2015). "Water Resources Allocation: Sharing Risks and Opportunities." https://read.oecd-ilibrary.org/environment/water-resources-allocation_9789264229631-en#page4 (accessed October 9, 2020).

PUB. (n.d.). "Water Price." https://www.pub.gov.sg/watersupply/waterprice (accessed October 9, 2020).

Ratnasiri, S., C. Wilson, W. Athukorala, M.A. Garcia-Valiñas, B. Torgler, and R. Gifford. (2018). "Effectiveness of Two Pricing Structures on Urban Water Use and Conservation: A Quasi-Experimental Investigation." *Environmental Economics and Policy Studies* 20, no. 3: 547–560.

Savenije, H.H.G. and P. Van Der Zaag. (2002). "Water as an Economic Good and Demand Management Paradigms with Pitfalls." *Water international* 27, no. 1: 98–104.

Soto Rios, C.P., A.T. Deen, N. Nagabhatla, and G. Ayala. (2018). "Explaining Water Pricing Through a Water Security Lens." *Water* 10, no. 9.

SSWM. (n.d.). "Subsidies (WD)." https://sswm.info/water-nutrient-cycle/water-distribution/softwares/economic-tools/subsidies-%28wd%29 (accessed October 9, 2020).

Texas Water Development Board. (2004). "Report 362 Water Conservation Implementation Task Force Water Conservation Best Management Practices Guide." http://www.twdb.texas.gov/publications/reports/numbered_reports/doc/R362_BMPGuide.pdf (accessed October 9, 2020).

Toronto Water. (n.d.). “Water Efficiency for Business.” https://www.toronto.ca/services-payments/water-environment/how-to-use-less-water/water-efficiency-for-business/ (accessed October 9, 2020).

UNECE. (2017). “Water Allocation in Transboundary Basins: Highlighting Global Relevance and Successful Cooperation.” https://www.unece.org/info/media/news/environment/2017/water-allocation-in-transboundary-basins-highlighting-global-relevance-and-successful-cooperation/doc.html (accessed October 9, 2020).

US EPA. (2011). “Water and Wastewater Pricing: An Informational Overview.” https://nepis.epa.gov/Exe/ZyNET.exe/901U1200.TXT?ZyActionD=ZyDocument&Client=EPA&Index=2000+Thru+2005&Docs=&Query=&Time=&EndTime=&SearchMethod=1&TocRestrict=n&Toc=&TocEntry=&QField=&QFieldYear=&QFieldMonth=&QFieldDay=&IntQFieldOp=0&ExtQFieldOp=0&XmlQuery=&File=D%3A%5Czyfiles%5CIndex%20Data%5C00thru05%5CTxt%5C00000011%5C901U1200.txt&User=ANONYMOUS&Password=anonymous&SortMethod=h%7C-&MaximumDocuments=1&FuzzyDegree=0&ImageQuality=r75g8/r75g8/x150y150g16/i425&Display=hpfr&DefSeekPage=x&SearchBack=ZyActionL&Back=ZyActionS&BackDesc=Results%20page&MaximumPages=1&ZyEntry=1&SeekPage=x&ZyPURL (accessed October 9, 2020).

US EPA. (n.d.). “Pricing and Affordability of Water Services.” https://www.epa.gov/sustainable-water-infrastructure/pricing-and-affordability-water-services#pricing (accessed October 9, 2020).

Wheeler, S.A., A. Loch, L. Crase, M. Young, and R.Q. Grafton. (2017). “Developing a Water Market Readiness Assessment Framework.” *Journal of Hydrology* 552: 807–820.

World Bank. (2019). “Doing More with Less – Smarter Subsidies for Water Supply and Sanitation.” https://www.worldbank.org/en/topic/water/publication/smarter-subsidies-for-water-supply-and-sanitation (accessed October 9, 2020).

WSSCC. (2009). “Public Funding for Sanitation – The Many Faces of Sanitation Subsidies.” https://www.wsscc.org/resources-feed/public-funding-sanitation/ (accessed October 9, 2020).

4

Water Augmentation

Introduction

Increased urbanization and changes in land use, along with climatic extremes, are affecting the natural cycle of aquifer recharge. Natural recharge occurs when surface water can percolate through the soil and vegetation to reach an underground water table that originates from rain or surface streamflow. With urban development and land conversion reducing the permeability of soils and thus limiting groundwater recharge, and groundwater recharge likely to decrease with climate change, water augmentation aims to increase the available supply of water through active recharge or protection of water recharge areas (CTCN 2017; Kahsay et al. 2018). This chapter will discuss a variety of water augmentation strategies, including conjunctive use and development of surface water and groundwater, managed aquifer recharge, and flood-managed aquifer recharge, as well as the role of source water protection.

Conjunctive Use and Development of Surface Water and Groundwater

In many parts of the world, high precipitation levels occur during a particular season of the year, which usually coincides with the lowest water demand. As such, water is typically transferred from the high-supply season to the high-demand season, with the most common solution being the storing of water behind dams. However, storing water in the ground may be a better alternative to surface storage systems, as surface reservoirs have many challenges, including evaporation and sedimentation. Furthermore, surface reservoirs have environmental impacts, including flooding of inhabited or good agricultural land, as well as high economic costs, with reservoirs requiring construction of costly canals due to the distance between dam and utilization area.

The United Nation's Food and Agricultural Organization (FAO) describes conjunctive use as harmoniously combining the use of both surface water and groundwater to minimize the undesirable physical, environmental, and economic effects of each solution and to optimize the water demand/supply balance. The practice can also help reduce the

Regional Water Security, First Edition. Robert C. Brears.

vulnerability of water supply to climatic extremes. Usually, conjunctive use of surface water and groundwater is considered within a river basin management program; specifically, both the river and the aquifer belong to the same basin (WRDMAP 2010; Zhang 2015).

There are two main phases of conjunctive use:

- *Recharge*: When the water table is high, the use of surface water is to be maximized. The recharge of groundwater can be enhanced artificially by surface water recharge and subsurface water recharge.
- *Recovery*: During the dry season, water is drawn from groundwater resources.

To keep recharge and recovery in balance, conjunctive water management involves the use of groundwater management tools, such as water balance estimation and water resource assessment, to help better manage water use between different parties, leading to well-prepared recharge and recovery phases and thus a lower probability of harming groundwater resources (SSWM 2019).

In order to use an underground reservoir to store significant volumes of water, possibly even of the same magnitude as the annual runoff, the groundwater reservoir should have sufficient free space between the ground surface and the water table to accommodate and retain the water to be recharged for the period when it is not needed. This requires accurate hydrogeological investigations, including geological mapping, geophysics, and reconnaissance drilling, to determine the configuration and the storage capacity of the underground reservoir. According to the FAO, the suitability of an aquifer for recharging may be estimated from the following parameters:

- The surface material must be highly permeable, to allow water to percolate easily.
- The unsaturated zone should present a high vertical permeability, and vertical flow of water should not be restrained by less permeable clayey layers.
- The depth to the water level should not be less than 5 meters.
- Aquifer transmissivity should be high enough to allow water to move rapidly from the mound created under the recharge basin but not so high that water cannot be recovered. An adequate transmissivity for recharge is also a good indicator of the aquifer's capacity to produce high well discharge and therefore easily return the water stored (FAO 1993).

Natural and Artificial Recharge

If a significant natural recharge of the aquifer occurs from the surface runoff and the deep percolation of rainfall, and if the average annual recharge amount is of the same magnitude as the water demand, then there is no need for any additional human intervention. Induced natural recharge occurs when intensive exploitation of groundwater close to a river results in a depression of the groundwater level and in a water inflow from the river. This is often seen in temperate climates where rivers flow all year round. Meanwhile, artificial recharge of aquifers can be achieved using three different methods:

- *Surface Spreading*: This consists in increasing the surface area of infiltration by releasing water from the source to the surface of a basin, pond, pit, or channel. This is an efficient and cost-effective method of aquifer recharge. However, only unconfined aquifers can be recharged this way, and large surface areas are required as they will allow water to evaporate if percolation in the ground is slow.
- *Watershed Management and Water Harvesting*: Watershed management is an effective method of intercepting dispersed runoff. Traditional terraced agriculture is one of the most common water harvesting methods in arid areas. Urban areas can also act as water harvesting systems, producing significant runoff that can be used for artificial recharge.
- *Recharge Wells*: Artificial recharge by injection consists in using a conduit access (e.g., a tubewell, shaft, or connector well) to convey water to the aquifer. This is the only method of artificial recharge of confined or deep-seated aquifers with poorly permeable overburden. The recharge is instant, and there is no transit or evaporation loss (FAO 1993; Ghazavi et al. 2018).

The conjunctive use of surface water and groundwater sources provides a variety of benefits, including:

- Greater water supply security, due to natural groundwater storage in aquifers.
- Larger net water supply yield than would typically be seen with one source alone.
- Better timing of irrigation water delivery, since groundwater can be rapidly deployed to compensate for any shortfalls in canal, tank, and other surface water availability at critical times in the crop growth cycle.
- Reduced environmental impact, through counteracting of waterlogging, salinization, excess river flow depletion, and aquifer over-exploitation.

To achieve the maximum water use benefits while minimizing associated environmental impacts, it is necessary to manage the conjunctive development of water resources such that:

- A balance is created between local recharge and groundwater use.
- The impact of surface water off-take does not extend into the low-river-flow period.
- The impact of waterwell abstraction is mostly delayed until high-river-flow periods (World Bank 2010).

Conjunctive Use of Surface Water and Groundwater for Irrigated Agriculture

The conjunctive use of surface water and groundwater for irrigated agriculture is beneficial as it can buffer the natural variability of supplies obtained from surface water by relying to a greater extent on groundwater (Schoups et al. 2006). A range of planning considerations should be taken into account in order to ensure effective management (Table 4.1) (World Bank 2010).

Table 4.1 Planning conjunctive use of surface and groundwater for irrigated agriculture.

Groundwater Parameter and Considerations	Management Measures and Instruments
Aquifer Characteristics: Storage capacity of shallow and intermediate-depth aquifers is critical	Surface water irrigation systems should avoid areas underlain by low-permeability strata, which are likely to result in soil waterlogging and salinization unless elaborate and costly drainage is provided
Recharge Excess in Local Groundwater Balance (Rising Groundwater Table and Soil Waterlogging): A balance needs to be struck between groundwater recharge and discharge by all mechanisms operative, and any "excess recharge" must be used productively	Surface water diversion can be curtailed so as to reduce excess recharge to groundwater Private waterwell development may be stimulated through a range of incentives Land reclamation and drainage measures may be required to kick-start intensive cultivation in a given area
Recharge Deficit in Local Groundwater Balance (Falling Groundwater Table and Saline Encroachment): Balance needs to be restored between groundwater recharge and discharge through waterwell use	Surface water resources may be diverted and reallocated through a corresponding reduction in direct groundwater use Recharge enhancement measures could be introduced, utilizing excess surface water flows through irrigation canals Less water-consuming crops could be promoted on the same irrigated area, to reduce consumptive water use
Groundwater Quality: Careful diagnosis of any incipient groundwater salinity problems in and around the area of irrigation development is essential	Selective irrigation canal lining could be implemented in areas with saline groundwater, to prevent further recharge and mobilization Supplementary surface water may be supplied in the case of coastal areas experiencing excessive groundwater abstraction and saline intrusion In some instances, saline groundwater might be put to a particular use (e.g., fish-farming ponds) as part of an overall plan for sustainable groundwater management

Source: World Bank, "Conjunctive Use of Groundwater and Surface Water: From Spontaneous Coping Strategy to Adaptive Resource Management". © 2010 The World Bank.

Case 4.1 The Sacramento Regional Water Bank

Water providers in the Sacramento region of California are developing the Sacramento Regional Water Bank, an innovative groundwater storage program that will improve regional water supply reliability in the near and long term. The Water Bank will operate by coordinating the use of surface water and groundwater. When surface water supplies are plentiful, water providers in the region will draw more water from Folsom Lake or local rivers and use it to offset existing demand for groundwater. This increases groundwater storage levels, resulting in a deposit in the Water Bank. During dry years, recovery of stored groundwater will occur through additional pumping, resulting in a withdrawal from the Water Bank. Regarding storage and recovery potential, the groundwater basin has more than 1.8 million acre-feet of available storage space. During a wet year, local water providers could store up to 60000 acre-feet of water in the Water Bank using existing facilities; this is enough water to serve more than 150000 households in the region for a year. Local water providers have also identified infrastructure improvements that, if implemented, could extend the Water Bank's capacity. Investing around $288 million in facility improvements (Table 4.2) could increase the region's storage and recovery capacities by more than 50% (Sacramento Regional Water Authority 2019a,b).

Table 4.2 Water bank infrastructure improvements.

Structural Actions	Contribution to Sacramento Regional Water Bank	Total Capital Cost (Estimated)
System interties	Installing interties between water systems could facilitate sharing of supplies among water providers in the Sacramento region. This would help distribute more surface water into the region in wet periods and allow the use of more groundwater during dry periods	$50 million
Groundwater well rehabilitation	Rehabilitating up to 16 existing groundwater wells could increase the volume of water that could be recovered from the Water Bank during dry years. Recovery operations would be carried out in accordance with local Groundwater Sustainability Plans	$16 million
New groundwater well installation	Installing up to 37 new groundwater wells could increase the volume of water recovered from the Water Bank during dry years. Recovery operations would be carried out in accordance with local Groundwater Sustainability Plans	$124 million
Groundwater injection and recovery wells	Installing or retrofitting up to 18 wells with both injection and recovery capability could store – or "deposit" – more water in the Water Bank during wet years for use during dry periods. These are commonly known as aquifer storage and recovery wells	$57 million
Booster pump/ pressure reduction	Installing booster pumps and pressure-reduction equipment could address pressure differences between water systems, improving the ability of water providers to share supplies with neighboring communities	$41 million
Total near-term infrastructure improvements		**$288 million**

Source: Sacremento Regional Water Authority, "Sacremento Regional Water Bank: A Sustainable Storage and Recover Program," (2019), https://rwah2o.org/wp-content/uploads/2019/04/WaterBank_Insert_9-FINAL.pdf; Sacramento Regional Water Authority, "Investing in Improved Regional Water Reliability," (2019), https://rwah2o.org/wp-content/uploads/2019/04/SRWB_Infrastructure_FactSheet_3.pdf.

Managed Aquifer Recharge

Managed aquifer recharge (MAR) consists of water management methods that recharge an aquifer using either surface or underground recharge techniques. The stored water is available for use in dry years when surface water supplies may be low. MAR can be used as a drinking water supply, as process water for industry, for irrigation, and for sustaining groundwater-dependent ecosystems. There are two approaches used to recharge aquifers: surface infiltration and deep injection. Surface infiltration can involve creating artificial streams and ponds (spreading grounds) in fast-draining soil, creating local catchment systems for rainwater and stormwater, or diverting water to naturally infiltrating river channels during low-flow seasons. Meanwhile, deep-injection methods put excess water directly into the aquifer using wells (American Geosciences Institute 2017).

MAR can be used to address a wide range of water management issues, including:

- Storing water in aquifers for future use.
- Smoothing out supply and demand fluctuations.
- Increasing resilience to droughts.
- As part of an integrated water management strategy.
- Stabilizing or raising groundwater levels when over-exploited.
- Situations where no suitable surface storage site is available.
- Reducing loss through evaporation and runoff.
- Impeding stormwater runoff and soil erosion.
- Improving water quality and smoothing fluctuations.
- Maintaining environmental flows in streams and rivers.
- Managing saline intrusion or land subsidence.
- Discharging treated wastewater and stormwater (Gale et al. 2009; Perrone and Rohde 2016; Dahlqvist et al. 2019).

Types of Managed Aquifer Recharge

There are a range of MAR types, as follows.

Infiltration Ponds

These are open-water ponds that are either excavated or constructed by directly flooding land surrounding a confining ditch with source water. They are installed in areas that have enough permeability and storage capacity for the target area. Also known as infiltration basins or percolation ponds.

Soil Aquifer Treatment

This involves reclaimed water percolating through an unsaturated zone, which is often several tens of meters thick. Residence times, usually several months, in the subsurface improve chemical and biological quality significantly.

Excess Irrigation

Excess irrigation can be applied in irrigated land where water is deliberately spread by, for example, sprinkler or drip irrigation during dormant or non-irrigated seasons to increase water availability during dry seasons. Ditches and trenches collect excess water, allowing it to infiltrate.

Reverse Drainage, Borehole, or Shaft Recharge

Boreholes or dug wells and shafts are used to recharge shallow phreatic aquifers where surface layers have low permeability and spreading methods are not feasible. Abandoned (dug) wells that have run dry due to falling water tables are often used for this purpose. During reverse drainage, the source water flows to a network of perforated drainage conduits and then infiltrates into the soil. This method is most appropriate where land is expensive, as it has minimal impact on surface land use.

Check Dams

Check dams are temporarily or permanently constructed dams built across stream channels to retain part of the stream discharge and enhance groundwater recharge. They are typically found in nonperennial streams. A release weir enables the discharge of ponded water in times of low infiltration and high evaporation to enhance infiltration downstream.

Riverbed Scarification

This is a method of enhancing recharge, extending the area through which infiltration takes place. The impermeable layer sedimented in the top of a riverbed is periodically removed to facilitate recharge through the alluvial materials.

Induced Bank Filtration

This involves the extraction of groundwater from a well near or under a river or lake to induce infiltration from the surface water body. Water quality improvement of induced surface water is often the main objective. Bank infiltration schemes usually consist of a well gallery or a line of abstraction wells parallel to the bank of a surface water body. They are typically installed near perennial streams and lakes that are in hydraulic contact with the adjacent aquifer.

Aquifer Storage, Transfer, and Recovery

Aquifer storage and recovery is the direct injection of water into a well for subsurface storage and later recovery; it is often used to bridge seasonal gaps in water supply, with water treatment being a secondary purpose. Aquifer storage, transfer, and recovery, on the other hand, is the injection of water into one well for storage followed by recovery from a different well. This is often done to extend the residence time in the aquifer and can be carried out continuously or intermittently, as needed. Direct injection is used when groundwater is deep, in terrains with hilly topography, or when existing land use makes surface spreading impractical or prohibitively expensive.

Aquifer Storage

Well injections are used to create a hydraulic freshwater barrier to counteract lateral seawater intrusion.

Sand Dams

Sand dams are constructed aboveground in a nonperennial riverbed. During periods of high flow, sand and gravel accumulate upstream of the dam. Runoff can easily infiltrate these highly permeable deposits, creating a small-scale artificial aquifer. Sand dams are a cost-effective (limited operation and maintenance costs), sustainable solution for rural water supply. Their lifespan is at least 30 years.

Subsurface Dams

Subsurface dams do not enhance groundwater recharge, but create additional storage by stopping groundwater flow upstream of constructed impermeable barriers. They require impermeable bedrock at a depth of a few meters (Page et al. 2018; DEMEAU n.d.).

Source Water

A variety of source waters may be used for underground storage, including surface water, groundwater, stormwater, treated wastewater, and even produced water. Water from different sources may have different quality characteristics. The water source used for recharge depends on availability, quality, duration, and reliability, as well as regulatory constraints. Variations in water quality and water quantity can be mitigated during storage provided there is adequate storage time and capacity. Water quality improvements can be made during pretreatment prior to recharge, during storage, and during post-treatment before use. Ideally, the selection of source waters will reduce the pre- and post-treatment requirements, since these increase overall system costs (National Research Council 2008).

Case 4.2 Groundwater Replenishment Scheme, Perth, Western Australia

The Groundwater Replenishment Scheme (GWRS) is a climate-independent water source for Perth. Commissioned in 2014 by the Government of Western Australia after successful completion of the 3-year Beenyup Groundwater Replenishment Trial, the 14 gigaliters/year (GL/year) GWRS is designed to treat recycled water from the Beenyup Wastewater Treatment Plant to a level of purity that is equal to or higher than drinking water quality and then inject it into the Leederville and Yarragadee aquifers for later abstraction for drinking water. The treatment process involves:

1) Secondary treated wastewater from Beenyup Wastewater Treatment Plant being pumped to the Advanced Water Recycling Plant for further treatment.
2) Water entering the Advanced Water Recycling Plant being treated through three barriers: ultrafiltration, reverse osmosis, and disinfection with ultraviolet light.
3) The water that has passed through the three barriers, now pure as drinking water, being injected into groundwater for storage.
4) The stored water being pumped out through a series of monitoring bores for sampling and testing.

The Water Corporation has now implemented Stage 2 of the GWRS, which will double the size of the scheme, with around 28 GL/year being recharged to the Leederville and Yarragadee aquifers at three locations. This will involve the construction of a second Advanced Water Recycling Plant at the Beenyup facility, drilling of new bores for recharge, and monitoring and building of a new 13 km-long pipeline connecting the Advanced Water Recycling Plant with the new bores. Overall, by 2060, groundwater replenishment could make up ≤20% of Perth's yearly drinking water supply (Government of Western Australia Department of Health 2016; Water Corporation 2018, 2020).

Source: Government of Western Australia Department of Health, "Groundwater Replenishment Scheme," https://ww2.health.wa.gov.au/Articles/F_I/Groundwater-replenishment-scheme; Water Corporation, "Groundwater Replenishment Scheme Stage 2 Expansion," https://www.watercorporation.com.au/water-supply/ongoing-works/groundwater-replenishment-scheme; "Groundwater Replenishment," (2020), https://www.watercorporation.com.au/-/media/files/residential/water-supply/ongoing-works-and-projects/groundwater-replenishment-scheme/groundwater%20replenishment%20brochure.pdf.

Flood Managed Aquifer Recharge

Flood managed aquifer recharge (flood MAR) is an integrated resource management strategy that uses flood water resulting from, or in anticipation of, rainfall or snowmelt for MAR on agricultural lands and working landscapes (e.g., refuges, floodplains, and flood bypasses). Flood MAR can be implemented at multiple scales, from individual landowners diverting floodwaters with existing infrastructures to extensive detention/recharge areas and modernizing flood management infrastructure/operations.

Benefits of Flood Managed Aquifer Recharge

In addition to enhancing water supply reliability, flood MAR provides a variety of benefits, including the following:

- *Flood Risk Reduction*: Aging infrastructure, deferred maintenance, and climate change have increased the flood risk to people and property. Flood MAR can reduce peak river flow and stage and reduce downstream flooding risk by diverting flows from river channels on to working landscapes. By implementing flood MAR, flood benefits are achieved in multiple ways, including:
  - Taking water off the channel during high-flow events (skimming peak flows) and purposefully delivering it to lands to promote groundwater infiltration. This requires flexibility and access to significant land area in order to achieve flood risk reduction benefits downstream of diversion points.
  - Lowering reservoir storage levels before, during, or after the flood season or discrete events, in order to reduce flood risks below the reservoir. The vacated water can be conveyed to specific areas for MAR.
  - Slowing runoff from properties to encourage groundwater infiltration on public and private lands and reduce runoff from entering already swollen channels.
- *Drought Preparedness*: Flood MAR results in more water being stored in aquifers. The stored water can be used for a variety of purposes, including in future dry years, droughts, and water shortages.
- *Aquifer Replenishment*: In many locations around the world, aquifers are in conditions of overdraft and groundwater levels continue to decline. Flood water is another source of water that can be used to replenish aquifers and reverse declining groundwater levels. Recharged groundwater can be extracted for beneficial use or left for the purpose of aquifer replenishment or to increase or maintain groundwater levels.
- *Ecosystem Enhancement*: Flood MAR can provide ecosystem benefits by reconnecting and inundating floodplains, creating floodplain habitat, marsh, and wetlands, supplementing baseflows, and supporting groundwater-dependent ecosystems through increased baseflow resulting from higher groundwater levels. Factors that determine the potential ecosystem and habitat enhancement opportunities are land use, proximity, and connectivity to the river, the timing of recharge flows, and the length of flooding. Seasonal flooding of land will boost food production to support aquatic and terrestrial species. Recharging groundwater supplies can also provide ecosystem benefits by increasing instream baseflow or reducing surface water temperature through surface water and groundwater interactions.
- *Subsidence Mitigation*: Flood MAR has the potential to reduce groundwater overdraft conditions and stop or slow land subsidence. Land subsidence can damage infrastructure,

including water supply conveyance facilities, levees, and flood channels, by reducing its water storage and carrying capacities. It also affects conveyance capacity by reducing hydrologic gradients. Land subsidence can permanently reduce the water storage capacity of an aquifer and even damage ecosystems if drainage is altered, leading to anoxic conditions.

- *Water Quality Improvement*: Flood MAR can improve groundwater quality by increasing the amount of water in storage and potentially diluting impaired or contaminated aquifers. Increasing water levels through flood MAR can also help prevent or slow seawater intrusion in coastal aquifers.
- *Working Landscape Preservation and Stewardship*: Flood MAR strategies can compensate landowners for keeping their land in its current use (e.g., agricultural production) while allowing periodic flooding. This allows farmland to stay in production, rather than retiring lands to create recharge basins. These recharge areas can be protected to ensure they remain available for recharge, rather than being converted to other uses (e.g., urban infrastructure).
- *Climate Change Adaptation*: Flood MAR improves the flexibility of water resource management systems to adapt to extreme events, which are expected to become more common in a changing climate. Flood MAR adds flexibility to system operations that need to compensate for lower snowmelt runoff and potential changes in water demand.
- *Recreation and Aesthetics*: Flood MAR has the potential to provide recreation and esthetic benefits based on the land's use of the recharge area. For example, increased flooding on refuges can improve bird watching and hunting and the creation of wetlands, while the reconnection of rivers and streams to floodplain habitats can enhance the beauty of a landscape compared with nonvegetated dedicated recharge basins (California Department of Water Resources 2018a,b).

Case 4.3 Potential of Flood MAR in the Merced River Watershed, California

The California Department of Water Resources, in partnership with the Merced Irrigation District, is conducting a preliminary study to determine the potential of flood MAR in the Merced River watershed. This is a proof of concept evaluating the feasibility, effectiveness, and benefits of planning and implementing a flood MAR project. The objective is to quantify the flood risk reduction benefits while analyzing the impacts of climate change. The study utilizes software from the US Army Corps of Engineers Hydrologic Engineering Center (HEC), with HEC-River Analysis System (HEC-RAS) models being used to estimate the flood stage, flow, and floodplain inundation of rivers and creeks and HEC-Flood Impact Analysis (HEC-FIA) to evaluate event-based damages. While HEC-FIA is traditionally used for this purpose, the watershed, reservoir, and groundwater models used to evaluate flood MAR span decades. The HEC-Watershed Analysis Tool software is used to run the HEC-RAS model. When a flood stage is triggered, its consequences are estimated using HEC-FIA for each flood event over the study period. The numbers of flood events and consequences by event are then used to determine the relative economic implications of the baseline conditions (without flood MAR actions) and three progressive "levels" of flood MAR actions based on a range of assumed reservoir and canal operations and conveyance and recharge capacities (Stygar et al. 2019).

Source: David Arrate Shem Stygar, Clark Churchill, and Lee Bergfeld, "Merced River Watershed Flood-Managed Aquifer Recharge Reconnaissance Study: Estimating Flood-Mar Consequences Using Hec-Ras and Fia," in AGU Fall Meeting (San Francisco 2019).

Source Water Protection

Source water refers to the surface water and groundwater supplies, such as lakes, rivers, and aquifers, that provide drinking water to humans. Source water protection involves implementing programs or activities targeted at reducing the likelihood of contaminants polluting these water resources and preventing their overuse. It consists in detecting threats associated with drinking water sources and enacting policies that protect water quality and water quantity. As such, source water protection is a proactive and substantial element of the successful water management of drinking water. It involves the implementation of a number of different activities to ensure the availability of satisfactory water quality and water quantity for various users and uses (Eledi 2016).

Benefits of Source Water Protection

The benefits of source water protection include reduced water treatment costs and increased water safety (protecting water at the source results in superior water quality, requiring less treatment). Furthermore, the costs of acquiring land in watersheds are lower than the investment required in systems for treating water. Source water protection also conserves limited water resources, particularly in rural communities (Eledi 2016). Other benefits include:

- Protecting biodiversity in upstream habitats.
- Erosion control and flood risk mitigation.
- Sustainable regulation of dry- and wet-season flows for upstream wetlands.
- Sediment regulation, enabling optimal operation of hydropower production facilities.
- Fostering collaboration between water utilities and farmers.
- Providing recreational opportunities and livelihood income opportunities (CTCN 2017; Mehan and Carpenter 2019).

Developing a Source Water Protection Plan

There are two main approaches to source water protection: controlling point source pollution and attaining standards for the management of diffuse pollution. All efforts should be focused on identifying possible pollution sources in the river basin to ensure that appropriate strategies and plans for source water protection are implemented.

A source water protection plan focuses on the threats and issues that lie outside a pumping station compound. The plan is an agreed statement between stakeholders that establishes objectives, actions, responsibilities, and funding. Successful plans will:

- Meet the individual needs of river basins.
- Bring all the key partners together to map out plans.
- Include critical technical data required to make the right decisions.
- Have the political support of all levels of government through funding, legislation, and regulations.
- Be available to river basin residents.
- Be adequately funded for development and ongoing updating.

The information frequently covered in source water protection plans includes water budgets, technical data, inventories of water resources and the river basin features that impact those resources, lists of current or potential sources of contamination, and lists of high-risk land uses and vulnerable areas (Conservation Ontario 2013). Source water protection plans can also include educational actions, such as developing outreach materials about source water protection, communicating with potential upstream threats, collaborating with other water utilities, and researching chemicals stored within the river basin (West Virginia Rivers 2017).

Regulatory Frameworks

For source water protection plans to be successful, regulatory frameworks should include the following:

- *Direct Regulations of Activities Affecting Source Water and Surrounding Land Use*: There are a variety of direct regulations that can be considered, including:
  - *Effluent Control*: This can involve setting targets, end-of-pipe control, and control of point sources of pollution.
  - *Control of Activities in Watersheds*: This can involve controlling the use of chemicals such as pesticides and fertilizers in river basins.
  - *River Basin Management Approach*: This can involve zoning and land-use planning.
  - *Codes of Practice*: These can be implemented to control land use and management activities (e.g., agricultural activities).
- *Financial Incentives and Disincentives*: Economic approaches to source water protection such as financial incentives/disincentives provide adequate support for regulatory mechanisms. In particular:
  - Charging for effluent is widely used and is based on the "polluter-pays principle."
  - Charging for water abstraction is another option.
  - Financial disincentives include fines for noncompliance with established permissible levels.
  - Subsidies and tax incentives can be used to encourage the purchasing of fertilizers and pesticides that pose a lower risk of source water contamination.
- *Evidence-Based Approaches*: Regulatory frameworks should be based on accurate data and current scientific knowledge. The World Health Organization's Guidelines for Drinking-Water Quality (WHO 2017) provide an objective and internationally accepted source of data for the development and implementation of health-based targets, qualitative microbial risk assessment, and quality measures for source water and drinking water. Assessing and monitoring the quality and quantity of water at intake points is a fundamental aspect of source water protection and should be included as a requirement in the regulatory framework.
- *Effective Institutional Support*: Developing and implementing source water protection plans requires effective institutions responsible for source water protection planning, implementation, and management. Institutional arrangements should be in place to assist with or assume these responsibilities. Examples of mechanisms for establishing such arrangements include memoranda of understandings and legally binding contractual agreements and performance contracts. As public health and environmental interests often overlap in source water protection, cooperation between public health and environmental authorities is important and should be pursued with priority.

- *Effective Regulations*: Source water protection regulations should be formulated in a way that ensures their effective implementation and enforcement. The roles and responsibilities of all stakeholders (both governmental and nongovernmental) and compliance measures (including the authority to ensure regulations are adequately enforced) should be clearly defined.

Policies and Programs

When developing policies and programs in support of source water protection plans, a range of factors need to be considered, as summarized in Table 4.3 (WHO 2011).

Nature-Based Solutions for Source Water Protection

Nature-based solutions (NBS) use nature's ability to capture, infiltrate, store, and filter water to ensure clean, reliable flows. Source water protection relies on a variety of NBS to protect the quality and quantity of drinking water at its source. The most common NBS-related activities for source water protection in upstream portions of river basins are listed in Table 4.4. Many source water protection programs employ multiple activities in parallel. In addition to addressing water security challenges, these activities have the potential to provide benefits to terrestrial and freshwater ecosystems (The Nature Conservancy 2017; Abell et al. 2019; Brears 2020).

Table 4.3 Factors to be considered in source water protection plans.

Focus of Policies and Programs	Description
Iterative involvement	Policies and programs should support the periodic review and update of source water protection plans to reflect, for instance, changes in the river basin or advances in scientific knowledge
Local context	Source water protection plans must be appropriate for the characteristics of and risks to specific water sources and catchment areas. As such, policies and programs should consider other aspects of local context, including the knowledge, attitudes, practices, and beliefs of local people
Stakeholder engagement	Stakeholders should be engaged in formulating and implementing source water protection plans. The views and needs of stakeholders should be identified, reviewed, and incorporated as they are formulated. Public and stakeholder involvement in the development of policies and programs can help to integrate and reflect these views
Stable and appropriate investment levels	Ensuring the appropriate level of investment, both financially and in terms of human resource capacity, is critical to ensuring that source water protection regulations and related policies and programs are implemented
Boundary waters	Because surface water and groundwater resources frequently cross administrative or political boundaries, boundary waters need to be considered when protecting source water, and appropriate legal instruments must be agreed to and applied

Source: Modified from WHO, "Protecting Drinking-Water Sources through Regulation," 2011.

Table 4.4 Nature-based solutions (NBS) for source water protection.

Activity	Description
Targeted land protection	Encompasses all the conservation activities undertaken to protect targeted ecosystems, such as forest, grassland, and wetland restoration initiatives. This reduces the loading of pollutants into waterways
Revegetation	Involves the restoration of natural forests, grasslands, and other habitats through planting (direct seeding) or by enabling natural regeneration. Revegetation restores the ability of nature to hold soil in place, reducing erosion, naturally filtering pollutants from overland flow, and helping infiltrate runoff water into the soil
Riparian restoration	Involves restoring natural habitat that is at the interface between land and water along the banks of waterways and lakes. These strips are commonly known as riparian buffers. They protect waterways from pollutants while providing shade, which can reduce water temperatures and improve dissolved oxygen content
Agricultural best management practices	Include practices such as cover crops, conservation tillage, and irrigation efficiency. In the context of source water protection, agricultural best management practices focus on land management practices that reduce erosion and nutrient runoff. These practices can help protect drinking water supplies and other uses, such as recreation and animal habitat
Ranching best management practices	Involve changes in land management practices on ranch lands that can create multiple environmental benefits. Ranching best management practices maintain or improve water quality and soils through a variety of methods, including the improvement of grazing management practices. These improvements reduce sediment and nutrient loadings and harmful pathogens from livestock waste
Fire risk management	Involves management activities that reduce forest fuels and hence the risk of catastrophic fire. Fire risk management seeks to achieve fuel reduction goals through mechanical thinning and controlled burns. This reduces the potential for catastrophic fires that abruptly remove forest cover and damage ground and cover soils, resulting in increased sediment and nutrient transport
Wetland restoration and creation	Involves either the reestablishment of the hydrology, plants, and soils of former or degraded wetlands that have been drained, farmed, or modified or the installation of a new wetland to offset wetland losses or mimic natural wetland functions. Wetlands improve water quality, provide habitat, store floodwaters, and maintain surface water flow during dry periods
Road management	The environmental impacts of roads include changes in hydrologic regimes and an increase in the number and extent of landslides and debris flows, all of which can affect water quality and terrestrial and aquatic systems. Mitigation actions include site-level actions to reduce erosion and improve road–stream crossings, the implementation of access management, and the closing and decommissioning of roads

Case 4.4 Source Water Protection in Bavaria, Germany

In Bavaria, 93% of drinking water comes from groundwater sources. Around 3500 water protection areas cover 4.5% of the state's territory. These are divided into three zones:

- *Protection Zone 1 "Fassungsbereich"*:
  - 10 meters' diameter around the well.
  - Any use is restricted.
  - Fenced.
- *Protection Zone 2 "Engere Schutzzone"*:
  - Diameter determined by 50-day flow duration.
  - Wastewater and manure are prohibited.
- *Protection Zone 3 "Weitere Schutzzone":*
  - The complete catchment of well.
  - Use of toxic materials is restricted.
  - Separated into Zones 3a and 3b if the catchment is vast.

To establish new source water protection areas, the following process is undertaken:

- The location and catchment of a well and the boundaries of its protection zones are determined (by hydrogeological engineering).
- The risks to the groundwater aquifer are determined (responsibility of utility).
- The concept is presented by the utility to municipal stakeholders and the municipal administration.
- A proposal for a new protection zone is submitted by the utility to the county or city administration (editing of proposal).
- The proposal is put on public display for 1 month + 2 weeks; comments can be submitted.
- A hearing is held, with invited experts and commentators.
- The permit is announced, including all relevant map materials (Ludwig 2016).

Source: Prof. Dr. Ralf Ludwig, "Water Supply and Source Water Protection in Munich, Germany: Status & Challenges," (2016), Retrieved from https://rv-eau.ca/wp-content/uploads/2016/11/Ludwig-Munich-Drinking-Water-Supply-Quebec-City-01-Nov-2016.pdf

References

Abell, R., K. Vigerstol, J. Higgins, S. Kang, N. Karres, B. Lehner, et al. (2019). "Freshwater Biodiversity Conservation through Source Water Protection: Quantifying the Potential and Addressing the Challenges." *Aquatic Conservation: Marine and Freshwater Ecosystems* 29, no. 7: 1022–1038.

American Geosciences Institute. (2017). "Managed Aquifer Recharge." Factsheet 2017-006. https://www.americangeosciences.org/geoscience-currents/managed-aquifer-recharge (accessed October 9, 2020).

Brears, R.C. (2020). *Nature-Based Solutions to 21st Century Challenges. Oxfordshire*: Routledge.
California Department of Water Resources. (2018a) "Flood-Managed Aquifer Recharge (Flood-MAR)." https://water.ca.gov/Programs/All-Programs/Flood-MAR (accessed October 9, 2020).
California Department of Water Resources. (2018b). "Flood-MAR: Using Flood Water for Managed Aquifer Recharge to Support Sustainable Water Resources." https://cawaterlibrary.net/document/flood-mar-using-flood-water-for-managed-aquifer-recharge-to-support-sustainable-water-resources/ (accessed October 9, 2020).
Conservation Ontario. (2013). "Everyone Lives Downstream." https://crca.ca/wp-content/uploads/PDFs/everyonelivesdownstream.pdf (accessed October 9, 2020).
CTCN. (2017). "*Climate Change Adaptation Technologies for Water: A Practitioner's Guide to Adaptation Technologies for Increased Water Sector Resilience*." CTCN Publications. https://www.ctc-n.org/resources/climate-change-adaptation-technologies-water-practitioner-s-guide-adaptation-technologies (accessed October 9, 2020).
Dahlqvist, P., K. Sjöstrand, A. Lindhe, L. Rosén, J. Nisell, E. Hellstrand, and B. Holgersson. (2019). "Potential Benefits of Managed Aquifer Recharge Mar on the Island of Gotland, Sweden." *Water* 11, no. 10.
DEMEAU. (n.d.). "Toolbox." https://demeau-fp7.eu/toolbox.html (accessed October 9, 2020).
Eledi, S. (2016). "Examining Source Water Protection Policies in Newfoundland and Labrador: A Literature Review." http://rplc-capr.ca/wp-content/uploads/2015/04/Examinng-SWP-Policies-in-NL-A-Literature-Review-3-mjb.pdf (accessed October 9, 2020).
FAO. (1993). "Land and Water Integration and River Basin Management." http://www.fao.org/3/v5400e/v5400e00.htm#Contents (accessed October 9, 2020).
Gale, I.N., Macdonald, D.M.J., and Calow, R.C. (2009). "Managed Aquifer Recharge: An Assessment of Its Role and Effectiveness in Watershed Management: Final Report for Dfid Kar Project R8169, Augmenting Groundwater Resources by Artificial Recharge: Agrar." http://nora.nerc.ac.uk/id/eprint/7453/ (accessed October 9, 2020).
Ghazavi, R., S. Babaei, and M. Erfanian. (2018). "Recharge Wells Site Selection for Artificial Groundwater Recharge in an Urban Area Using Fuzzy Logic Technique." *Water Resources Management* 32, no. 12: 3821–3834.
Government of Western Australia Department of Health. (2016). "Groundwater Replenishment Scheme." https://ww2.health.wa.gov.au/Articles/F_I/Groundwater-replenishment-scheme (accessed October 9, 2020).
Kahsay, K.D., S.M. Pingale, and S.D. Hatiye. (2018). "Impact of Climate Change on Groundwater Recharge and Base Flow in the Sub-Catchment of Tekeze Basin, Ethiopia." *Groundwater for Sustainable Development* 6: 121–133.
Ludwig, R. (2016). "Water Supply and Source Water Protection in Munich, Germany: Status & Challenges." https://rv-eau.ca/wp-content/uploads/2016/11/Ludwig-Munich-Drinking-Water-Supply-Quebec-City-01-Nov-2016.pdf (accessed October 9, 2020).
Mehan, G.T. III and A.T. Carpenter. (2019). "Bringing Agriculture and Drinking Water Utilities Together for Source Water Protection." *Journal – AWWA* 111, no. 8: 34–39.
National Research Council. (2008). "Prospects for Managed Underground Storage of Recoverable Water." https://www.nap.edu/catalog/12057/prospects-for-managed-underground-storage-of-recoverable-water (accessed October 9, 2020).

Nature Conservancy, The. (2017). "Beyond the Source." https://www.nature.org/en-us/what-we-do/our-insights/perspectives/a-natural-solution-to-water-security/ (accessed October 9, 2020).

Page, D., E. Bekele, J. Vanderzalm, and J. Sidhu. (2018). "Managed Aquifer Recharge (MAR) in Sustainable Urban Water Management." *Water* 10, no. 3.

Perrone, D. and M.M. Rohde. (2016). "Benefits and Economic Costs of Managed Aquifer Recharge in California." *San Francisco Estuary and Watershed Science* 14, no. 2.

Sacramento Regional Water Authority. (2019a). "Investing in Improved Regional Water Reliability." https://rwah2o.org/wp-content/uploads/2019/04/SRWB_Infrastructure_FactSheet_3.pdf (accessed October 9, 2020).

Sacramento Regional Water Authority. (2019b). "Sacremento Regional Water Bank: A Sustainable Storage and Recover Program." https://rwah2o.org/wp-content/uploads/2019/04/WaterBank_Insert_9-FINAL.pdf (accessed October 9, 2020).

Schoups, G., C.L. Addams, J.L. Minjares, and S.M. Gorelick. (2006). "Reliable Conjunctive Use Rules for Sustainable Irrigated Agriculture and Reservoir Spill Control." *Water Resources Research* 42, no. 12.

SSWM. (2019). "Conjunctive Use." https://sswm.info/sswm-university-course/module-4-sustainable-water-supply/further-resources-water-sources-hardware/conjunctive-use (accessed October 9, 2020).

Stygar, S., D. Arrate, C. Churchill, and L. Bergfeld. (2019) "Merced River Watershed Flood-Managed Aquifer Recharge Reconnaissance Study: Estimating Flood-MAR Consequences Using HEC-RAS and FIA." *AGU Fall Meeting*, San Francisco, CA.

Water Corporation. (2018). "Groundwater Replenishment Scheme Stage 2 Expansion." https://www.watercorporation.com.au/Outages-and-works/Ongoing-Works/Groundwater-Replenishment-Scheme-Stage-2-Expansion (accessed October 9, 2020).

Water Corporation. (2020). "Groundwater Replenishment." https://www.watercorporation.com.au/Our-water/Groundwater/Groundwater-replenishment (accessed October 9, 2020).

West Virginia Rivers. (2017). "Source Water Protection Plan Implementation: For Utilities, Watershed or Other Community Groups." https://wvrivers.org/wp-content/uploads/2017/05/SWPP-implementation-guide-FINAL4.12.17.pdf (accessed October 9, 2020).

WHO. (2011). "Protecting Drinking-Water Sources Through Regulation." https://www.who.int/water_sanitation_health/water-quality/regulation/sheet4.pdf?ua=1 (accessed October 9, 2020).

WHO. (2017). "Guidelines for Drinking-Water Quality, 4th Edition, Incorporating the First Addendum." https://apps.who.int/iris/bitstream/handle/10665/254637/9789241549950-eng.pdf;jsessionid=0C049FD2BC95D4D0C0DF4CB9C0932D30?sequence=1 (accessed October 9, 2020).

World Bank. (2010). "Conjunctive Use of Groundwater and Surface Water: From Spontaneous Coping Strategy to Adaptive Resource Management." http://documents.worldbank.org/curated/en/874731468315319173/pdf/575560replacem1onjunctive1Use120100.pdf (accessed October 9, 2020).

WRDMAP. (2010). "Conjunctive Use of Groundwater and Surface Water." Thematic Paper 2.6/3. https://assets.publishing.service.gov.uk/media/57ab2f6740f0b608a7000061/TP2.6.3-Conjunctive_use-rev04FINAL.pdf (accessed October 9, 2020).

Zhang, X. (2015). "Conjunctive Surface Water and Groundwater Management under Climate Change." *Frontiers in Environmental Science* 3, no. 59.

5

Water Efficiency

Introduction

Increased demand for water due to population growth, climate change, and growth in agriculture and industry is driving the need to use water resources more efficiently. There is also political pressure on water utilities to constrain price inflation on water bills, as well as legislative requirements to ensure there is enough water in the environment to protect aquatic flora and fauna (CIWEM 2016). Improving water efficiency reduces water scarcity and maximizes the benefits provided by existing water infrastructure. It also frees up water for other users, both human and natural (Brears 2020). This chapter will discuss a range of water efficiency measures aimed at achieving water security, including water metering, leakage reduction, public education, skill-base enhancement, and water efficiency labeling, as well as water efficiency measures in buildings, industrial facilities, and agricultural production.

Water Metering

Automatic meter readers (AMRs) are "one-way" readers that send usage data back to the utility. In comparison, advanced metering infrastructure (AMI) is a "two-way" solution that creates a network between meters and a utility's information system. Data flows both ways, facilitating not only remote meter reading but also, for example, enhanced conservation awareness programs. Specifically, AMI smart meters target the management and extraction of useful information from large amounts of high-resolution consumption data. The data can be used to develop customized awareness programs that can influence behavior change. From the utility side, smart meters provide many benefits, including leakage detection/energy reduction, demand forecasting, enhanced conservation awareness campaigns, promotion of efficient appliances, variable water pricing/tariffs, arguments for investment, and performance indicators. On the customer side, benefits include:

- Comparison of one's water consumption with other consumers (e.g., with neighbors in the same building or street).
- Comparison of one's water consumption with standard profiles (e.g., consumers with the same sociodemographic factors).

Regional Water Security, First Edition. Robert C. Brears.

- Comparison of one's water consumption with the most efficient users.
- Comparison of energy patterns associated with water use in the same household.
- Information on specific and alternative pricing schemes.
- Forecasting of the next water bill.
- Forecasting of the component of the next energy bill associated with water consumption (Brears 2016).

Case 5.1 The City of Houston's Water Usage Alerts

The City of Houston's water utility customers can sign up for usage alerts. High bills, resulting from increased registered water usage, are the responsibility of the customer, irrespective of whether they have access to these alerts. Customers who sign up for high usage alerts receive a notification if their water usage is higher than usual. If there is a broken pipe or leaking toilet, a high usage alert could be the first indicator of a potentially large water bill. Customers that sign up to the service can:

- Receive usage alerts and notifications via email.
- Access usage information via computer or phone.
- View their water usage by monthly, weekly, daily, and hourly patterns.
- Track their use history.

Customers will also receive alerts according to their set threshold via email, informing them of the following:

- Current gallon consumption.
- Dates when this consumption was used.
- Constant flow detection (which helps to identify leaks and other high bill causes) (Houston Public Works n.d.).

Source: Based on Houston Public Works, "High Usage Alerts," https://www.houstonwaterbills.houstontx.gov/ProdDP/Default/Default.

Reducing Unaccounted-For Water

A major issue for water utilities is addressing the considerable difference between the volume of water they treat and distribute and the volume for which the customer is invoiced. This gap is known as unaccounted-for water (UFW), and often amounts to between 25 and 50% of the total volume of water collected, treated, and distributed. The three main categories of UFW are apparent (or commercial) losses, caused by inaccurate customer metering, data-logging errors, and illegal connections to the network; real (or physical) losses, comprising leakage from all parts of the system; and overflow at the utility's storage tanks. Real losses are caused by poor operations and maintenance, lack of an active leak-detection program, poor quality of underground assets, or unbilled authorized consumption by the utility for operational purposes (e.g., water used for firefighting or provided free to specific customer groups). High levels of UFW impact the financial viability of water utilities due to revenue losses and unnecessarily high operating costs. UFW also impacts

Table 5.1 Benefits of reducing unaccounted-for water.

Benefit	Description
The same amount of water for more people	Reducing UFW will at least postpone the need for additional water resources in cities with a growing population. Up to 30% more people can be potentially served by making distribution systems more efficient. Otherwise, the growing demand for water in cities will increase the cost of water treatment and pumping, requiring extra plant capacity and more raw water. Also, expanding the water distribution network without a leak-detection program is effectively expanding a cycle of inefficiency
Lower operational costs	UFW means loss of not only water but also the energy used to treat and distribute it. Energy savings can be obtained by reducing leakage, since leakage-reduction programs ensure more stable water pressure through the system, increasing energy efficiency. Also, reduced pressure and less fluctuation will extend the life expectancy of pipes and valves
Higher revenues	High levels of UFW caused by inaccurate metering impact the financial viability of water utilities because of lost revenues – typically up to one-quarter of UFW comprises apparent (commercial) losses
Safe water quality	In most cities, it is necessary to add chlorine to disinfect water before distribution. If the distribution system does not work correctly, however, this will result in excessively high chlorine content nearer to pumping stations and low or zero chlorine further out. The flow and age of the distributed water need to be as uniform as possible over the entire system to make chlorination efficient. If the chlorine degrades, a distribution system with high leakage is vulnerable to contamination resulting from vacuums developing during low-pressure situations

Source: Brears, Urban Water Security; Qiang Xu et al., "Review on Water Leakage Control in Distribution Networks and the Associated Environmental Benefits," Journal of Environmental Sciences 26, no. 5 (2014); Nourhan Samir et al., "Pressure Control for Minimizing Leakage in Water Distribution Systems," Alexandria Engineering Journal 56, no. 4 (2017).

the capacity of utilities to fund necessary maintenance of the system. UFW can be reduced by improving leak-detection capability through a combination of predictive modeling and monitoring instruments, optimizing pipe maintenance strategies, and developing an instant pressure regulation system. Reducing UFW provides a range of benefits, as listed in Table 5.1 (Xu et al. 2014; Brears 2016; Samir et al. 2017).

Case 5.2 Detecting Water Leaks from Space in Victoria, Australia

The Victorian Government has completed a pilot program using satellite scanning to detect water leaks underground. Up to 160 million liters of drinking water are lost to leaks in Victoria every day, equivalent to 12% of the total annual water revenue for the state. The program involved Victoria's 17 water corporations and VicWater partnering with a private-sector company, Utilis, to verify whether leaks could be detected through scans from space. A 3000 square kilometer area of north Melbourne was scanned by a Japanese Aerospace Exploration Agency-operated satellite orbiting around 650 kilometers above ground. The Utilis system used microwave reflectometry to detect a spectral signature that matches drinking water. Aerial images from the satellite were overlaid on geographic information system maps to identify where the leaking pipes were located and how big the leaks were, and field crews were dispatched to repair them. Leaks as small as 85–246 milliliters per minute could be detected. During the pilot, the Utilis system found 18 leaks in different Melbourne suburbs in 4 days (ITNews 2016).

Source: Based on ITNews, "Victoria Trials Satellite Scans to Detect Water Leaks from Space," https://www.itnews.com.au/news/victoria-trials-satellite-scans-to-detect-water-leaks-from-space-416960. © nextmedia Pty Ltd

Public Education and Awareness

Education of the public is crucial in generating an understanding of water scarcity and creating acceptance of the need to promote water efficiency-related programs. Water utilities can raise public awareness through many formats, including:

- *Public Information*: This can include printed literature, public service announcements, advertisements on billboards and public transportation, television commercials, newspaper articles and advertisements, Internet and social media campaigns, and water utility websites.
- *Public Events*: This can include conservation workshops, expos, fairs, and citizen science projects.
- *Information in Water Bills*: Water bills should contain water conservation tips to help customers make informed decisions on future water use (Hoy and Stelli 2016; Hove and Osunkunle 2019; Seelen et al. 2019; Brears 2020).

Case 5.3 San Diego County Water Authority Advancing Outdoor Water Use Efficiency

The San Diego County Water Authority has developed resources to help homeowners and landscape professionals increase water efficiency countywide and make WaterSmart living a permanent way of life. These include a series of online videos to help guide homeowners through the process of transforming their property into a water-efficient landscape and the local introduction of a US Environmental Protection Agency (EPA)-approved training program for landscape professionals. More specifically:

- *Online Videos*: The video series offers in-depth instruction for upgrading turf yards into water-efficient landscapes that will thrive in the San Diego region. It consists of 17 short episodes, beginning with an introduction to water issues in the San Diego region and transitioning into six steps toward implementing a successful landscape makeover, including identifying water-efficiency targets, creating a basic plot plan, evaluating sites and soils, and implementing plans.
- *Training*: The San Diego County Water Authority is a partner in the Qualified Water-Efficient Landscaper (QWEL) training program. This program, which meets EPA WaterSense standards, provides landscape professionals with 20 hours of instruction, covering lessons about proper plant selection for local climates and irrigation

system design, maintenance, programming, and operation. Participants receive a certificate after demonstrating their ability to complete an irrigation system audit and passing a QWEL exam. Successful participants are also listed on QWEL's national directory for trained professionals (San Diego County Water Authority 2016).

Source: San Diego County Water Authority, "Water Authority Unveils Resources to Advance Region's Outdoor Water-Use Efficiency," https://www.sdcwa.org/water-authority-unveils-resources-advance-regions-outdoor-water-use-efficiency.

School Education

Schools use large volumes of water each day for heating and cooling systems, toilets, water faucets, cafeterias, laboratories, outdoor playgrounds, swimming pools, and landscaping. Water efficiency programs provide an opportunity for students to regard water conservation as the norm, making it more likely that there will be a change in how water is used across society. Water efficiency initiatives can also become part of the school curriculum, with students not only seeing the process of conservation and saving water in action but also participating in helping to collect data on the savings achieved through the installation of new water-efficient devices. Students can also participate in field trips that enhance awareness of the water cycle, such as trips to water treatment plants and waterways. Integrating water efficiency into school life not only benefits the environment but also provides a range of economic and social benefits, including:

- *Decreasing Water and Energy Costs*: Water efficiency decreases a school's water and energy costs, freeing up funds for its budget.
- *Building Young Leaders and Providing Learning Opportunities*: Water efficiency activities provide leadership chances and practical learning activities for students, plus occupational development for staff.
- *Establishing a Strong School Culture Based on Good Communication and Shared Goals*: Getting a whole school to work together allows the school, staff, and students to share in success.
- *Raising a School's Profile*: When a school participates in water efficiency activities, it relates to the community through partnerships and local networks, increasing the school's reputation in terms of environmental protection.
- *Contributing to Attaining a Better Environment through Adopting Water Efficiency*: Water efficiency activities allow a school to actively contribute to a more sustainable planet for current and future generations (Amahmid et al. 2019; El-Nwsany et al. 2019).

Case 5.4 Actsmart Schools in the Australian Capital Territory

Actsmart Schools is a free specialized program that helps schools improve their sustainability. Every school in the Australian Capital Territory has now signed up to the program and begun saving energy, conserving water, increasing recycling, protecting biodiversity, reducing greenhouse gas emissions, and integrating sustainability into the curriculum. The program provides a range of support services, including best practice guidelines, site visits, advice from specialized staff, professional learning, tools, and resources. Schools are evaluated through an accreditation system and rewarded for their efforts. Regarding water, Actsmart Schools provides the following resources:

- A Water Best Practice Guide, with step-by-step instructions and recommendations to help schools reduce water consumption and develop a Water Action Plan.
- Water links to the Australian Curriculum.
- Professional learning in water efficiency and water quality for teachers, business managers, and facilities managers.
- Examples of best practice and case studies on sustainable water usage.
- Visits from Actsmart Schools Officers to discuss current situations and an approach to water-focused behavior change activities.
- An analysis of water usage compared to benchmark levels (ACT Government n.d.).

Source: ACT Government, "Water Resources," https://www.actsmart.act.gov.au/what-can-i-do/schools/actsmart-schools/water/resources.

Developing a Resource-Efficiency Skill Base

Post-secondary education and training for resource efficiency provide opportunities for lifelong learning about the environmental gains and economic opportunities associated with resource efficiency. A range of education and training opportunities are listed in Table 5.2 (UNIDO 2011).

Table 5.2 Education and training opportunities to enhance resource efficiency.

Opportunity	Description
Technical and vocational training	Training schemes provide knowledge, methods, and tools to help students design and implement practical solutions to resource efficiency as they enter the workforce
Higher education	Universities prepare society's future decision-makers for their entry into the labor market. This includes training of future leaders, who will play essential roles in providing education at both the primary and the secondary levels
Lifelong/on the job	Education and training for resource efficiency are continuous processes that can be offered to employees and to society at large. For example, training measures can be applied to employees to raise awareness about resource efficiency potentials

Source: Modified from UNIDO, "Unido Green Industry: Policies for Supporting Green Industry," (2011), https://www.unido.org/sites/default/files/2011-05/web_policies_green_industry_0.pdf.

Case 5.5 Singapore's Water Efficiency Manager Course

Jointly developed by Singapore's Public Utilities Board (PUB) and Singapore Polytechnic (SP), the Water Efficiency Manager Course equips facilities managers with the knowledge and skills to conduct water audits and apply water efficiency measures. This leads to lower water consumption in commercial/residential and industrial areas. The course is aimed at building owners and facilities, estates, and operations managers, as well as engineers and architects. It runs for 3 days, and successful participants receive a joint PUB-SP Certificate of Proficiency. The course contents are detailed in Table 5.3 (PUB 2015).

Table 5.3 Singapore's Water Efficiency Manager Course content.

Day	Topic	Content
1	Introduction	• Regulatory water efficiency requirements • Incentives/recognition schemes • Guidelines for buildings
	Water Efficiency Management Plan, Audit Methodology, and Tools	• Audit process • Conducting an audit • Water Efficiency Management Plan (practical)
2	Cooling Towers	• Basic design/regulatory requirements • Water management • Best practice guidelines/case studies
	Water Efficient Irrigation and Landscaping	• Methods • Best practice guidelines/case studies
	Swimming Pools	• Leakage • Balancing tanks • Best practice guidelines/case studies
3	Industrial Water Solutions	• Conventional process/membrane treatment process • Overview of wastewater characteristics and common treatment/reclaim technologies • Water balance chart • Optimal water management of planning and construction plants • Case studies Site visit Test

Source: Singapore's Water Efficiency Manager Course, https://www.pub.gov.sg/Documents/WEMC2015.pdf

Water Labeling

Eco-labels and certification encourage sustainable consumption by providing consumers with information about the environmental impact of products and services. Companies are also rewarded for designing products and services that have the least possible environmental impact, further encouraging attitudes toward more pro-environmental management. Eco-labels and certifications can apply to the entire life cycle of a product or to a specific stage or step in the production process. They are generally awarded by a third party that authorizes the use of the label/certification on products in a particular product category (Brears 2018). In the context of water resources management, the labeling of appliances according to water efficiency is essential in reducing water consumption by eliminating unsustainable products from the market. However, this assumes that the labeling scheme is clear and understandable and identifies both private and public benefits of conserving water. People are more likely to respond to eco-labels if the environmental benefits match closely with personal benefits such as reduced water bills (Brears 2016).

Case 5.6 Australia's Water Efficiency Labelling and Standards Scheme

The Water Efficiency Labelling and Standards (WELS) scheme is designed to help consumers choose more water-efficient products for their home or business. It is estimated that water-efficient products could save Australia more than $2 billion by 2030, with 65% of the savings attributed to reduced electricity and gas costs from avoided water heating and the remaining 35% to reduced water bills. Products regulated under the WELS scheme are plumbing fixtures (faucets, showers, flow controllers, toilets, and urinals) and whitegoods (clothes washing machines and dishwashers). All must be registered with the WELS Regulator and labeled with the correct water rating information:

- *Star Rating*: This allows consumers to quickly compare the water efficiencies of different products (the more stars, the more water efficient a product is).
- *Rate of Water Consumption*: This could include flow rate, liters per flush, or liters per wash, helping customers estimate how much water the product will use.
- *Registration and Product Details*: These include license and registration numbers and, for washing machines and dishwashers, the model name and cycle used for testing (Australian Government Water Rating 2017).

Source: Modified from Australian Government Water Rating, "Choosing Water Efficient Products," https://www.waterrating.gov.au/choose

Water Efficiency in Buildings

Buildings, including residential buildings and institutional and commercial buildings, such as schools, hospitals, shopping centers, office buildings, hotels, and restaurants, are places where significant amounts of water are consumed. Water efficiency programs in buildings require preparation and planning, including conducting a facility survey, monitoring use, determining performance targets, identifying savings options, informing and engaging building users, and allocating resources. Examples of questions that can facilitate a site survey for a building include:

- How much water is consumed in different functions of the building?
- Which functions are the primary consumers: heating, ventilation, and air conditioning (HVAC), toilet facilities, technical areas, irrigation, others?
- What are the direct and indirect costs of water use in the building?

- What type of maintenance routine (e.g., leak inspection, equipment maintenance) is in place today?
- What water reuse and recycling systems are in use or have been considered?

Water Efficiency Measures for Buildings

There are many technologies that can be employed to reduce the amount of water used in a building.

Education Programs

One of the first steps toward reducing water use in a building is to train employees on the use of new water-efficient technologies and to train maintenance staff on their operation and maintenance. Technology that is not correctly used or maintained will fail to achieve its maximum savings potential.

Audits, Leak Detection, and Repair

Performance of water distribution system audits is not limited to existing, aged buildings. Water leaks can occur in newly constructed buildings due to improper installation of piping and fixtures or impacts to piping systems during the construction process. Therefore, during a building's design, a water budget should be developed and a water distribution system should be commissioned. Post-construction, if water consumption is greater than 10% of the baseline water budget estimated during the design process, a leak detection survey may be necessary. Water system leaks should be repaired immediately as they can damage building structures and increase water bills.

Toilets and Urinals

Toilet and urinal flushing generally accounts for one-third of a building's total water consumption. Significant water savings can be achieved through the replacement of older models of toilets and urinals with newer and more water-efficient ones:

- *Low-Volume or Dual-Mode Flush Systems*: Traditional toilets use more than 11 liters of water per flush. Modern, low-volume, dual-mode flush systems can reduce this to 4.5 liters per full flush and 3 liters per partial flush.
- *Vacuum Toilets*: Toilets can be connected to a vacuum source for flushing. These systems operate with the assistance of a pump that creates a vacuum to help flush the contents of the toilet with minimal water use.
- *Urinals with On-Demand Sensors*: Infrared sensor-operated urinals work by detecting the presence of a user within the detection zone for more than a certain time threshold. The user's departure from the detection zone activates flushing.
- *Waterless Urinals*: Waterless urinals have a drain trap insert siphon that collects the urine and discharges it into the sewage system, without using water.

Showerheads and Faucets

A range of low-flow showerheads can be installed, including atomizers, which deliver water in small but abundant droplets that cover larger surface areas; pulsators, which vary spray patterns with pauses between spurts or by pulsating between strong flow and light mist; and aerators, which mix water droplets with air to cover the desired surface area. Low-flow faucets employ aeration technology and sensors to achieve water savings at equivalent comfort levels to older, more consumptive fixtures. Examples of newer low-flow faucet technologies include metered valve faucets, which deliver a preset amount of water and then automatically shut off; self-closing faucets, which are spring-loaded so that they shut off a few seconds after being turned on; and ultrasonic and infrared sensor faucets, which automatically activate when hands are detected beneath them and shut off again when they are removed.

Boiler and Steam Systems

Strategies for achieving water and cost savings from boiler and steam systems include:

- Properly sizing boiler and steam units (and possibly downsizing units by reducing facility heat and steam requirements, lowering water makeup requirements).
- Specifying the most cost-effective system across the life cycle, and not just the one with the lowest upfront cost.
- Installing a smaller summer boiler or distributed system for reheat or dehumidification requirements instead of inefficiently running a larger boiler at part load.
- Considering alternative technologies such as heat pumps for small loads (Liu and Ping 2012; Arab Forum for Environment and Development 2014; WBDG 2016; Tarhule 2017).

Water-Efficient Landscaping

Landscapes can be made water-efficient through designs that require minimal supplemental water and the installation and maintenance of an irrigation system that efficiently applies water in the appropriate amount. For a list of recommended practices for building owners and facility managers looking to implement a water-efficient landscape, see Table 5.4 (US EPA 2017).

Case 5.7 Energy and Water Use Benchmarking in Los Angeles

On June 1, 2019, the Existing Building Energy and Water Efficiency Ordinance went into effect for the City of Los Angeles. The ordinance requires that all existing nonresidential buildings in the city over 20 000 square feet annually submit benchmarking data for water and energy use through the ENERGY STAR Portfolio Manager, and ideally plan to undertake initiatives that reduce energy and water usage in their facilities. Through the implementation of the ordinance, the city is planning to diagnose energy-inefficient buildings and aid in the implementation of energy-saving features. Once data are reported for a building, a cycle will begin whereby every 5 years, either prescriptive or performance measures are undertaken to improve its overall energy efficiency. These measures might involve energy and water audits, retro-commissioning, or energy and water reduction targets (Engineering Economics n.d.).

Source: Inc Engineering Economics, "Los Angeles Existing Buildings Energy and Water Efficiency Ordinance," http://www.eeiengineers.com/los-angeles-existing-buildings-energy-and-water-efficiency-ordinance/. © Engineering Economics, Inc.

Table 5.4 Recommended practices to increase outdoor water efficiency.

Practice	Description
Landscaping agreements	Periodically review all landscape service and maintenance agreements to incorporate a high priority for water, chemical, and energy efficiency
Hiring practices	Consider hiring landscape contractors who focus on water-efficient or climate-appropriate landscaping
Reporting and fixing issues	Require landscape contractors to report and fix problems. Many landscapers, in addition to installing and maintaining plants, will install and maintain an irrigation system. As such, they should identify and report leaks and other inefficiencies in the landscape
Use of mulch	Add mulch to plant beds to decrease water loss from the soil through evaporation
Use of sufficient topsoil	Maintain enough topsoil rich in organic matter (4–6 inches) to capture stormwater as it falls and release it back to plants over time, reducing irrigation requirements
Aeration of soils	Aerate soil seasonally in turf areas and landscape beds to introduce oxygen into its deep layers and break up compacted soil
Recirculation of water in features	Recirculate water in decorative fountains, ponds, and waterfalls. These features should be shut off when possible to reduce evaporative losses. The water recirculation systems should be checked annually for leaks and other damage. Nonpotable water should be considered for use in these systems, too

Source: Based on US EPA, "Water Efficiency Management Guide Landscaping and Irrigation ", (2017), https://www.epa.gov/sites/production/files/2017-12/documents/ws-commercialbuildings-waterscore-irrigation-landscape-guide.pdf

Industrial Water Efficiency

In industrial facilities, water is used in a wide range of activities, including: incorporation into final products; washing or rinsing of raw materials, intermediates, or final products; preparation of solvents or slurries; cleaning of equipment and space; removal or provision of heat; meeting of domestic and hygienic needs; and irrigation of landscape space. Increasing water efficiency in industrial facilities requires identification and measurement of quantitative and qualitative parameters for the water used and lost, as well as for any effluents generated. A checklist of questions to consider runs as follows:

- How much water is entering the process/area?
- What is the quality of the water at the point of entry?
- What is the cost of bringing water to this process/area?
- How much of the water entering this process/area is being incorporated into the final product?
- How much is rejected with effluent streams?
- What is the cost of managing the effluents?

Table 5.5 Measures to improve industrial water efficiency.

Measure	Description	Examples
Improved production planning and sequencing	Readjusting the production plans with a focus on minimizing water consumption	Reducing cleaning needs by minimizing product changes
Good housekeeping	Introducing more water-wise routines in operations	Avoiding spillages Minimizing the transport of pollutants from one process to the next Performing mechanical cleaning before washing with water
Process/equipment modifications	Making modifications in processes or equipment with relevant retrofits	Closing open-ended cooling or heating systems Installing level-controlled valves to avoid overflows Installing self-shutting, trigger-controlled nozzles on hoses
Product/material changes	Changing feedstocks used in production or designing entirely new products that lead to reduced water demand or less effluent generation	Switching to water-based paints Using reactive dyes in textile dyeing
Replacement of equipment/ technology	Substituting existing technologies with more effective and efficient ones	Adopting in-place cleaning systems Using high-pressure, low-volume cleaning equipment

Improving Industrial Water Efficiency

There are many options for improving industrial water efficiency. Table 5.5 lists some of the conventional ones, in order of increasing complexity of process change (Arab Forum for Environment and Development 2014).

Case 5.8 Welsh Water's Business Service: Process Audit Service

Welsh Water provides business customers with a process audit service whereby it will help reduce their overheads by lowering their water and energy bills and at the same time reduce their carbon emissions. This process audit is carried out by a Welsh Water industrial process engineer, who will assess the organization's water consumption within the production and operational processes on-site. The service can be scaled to reflect the needs of the customer, but the key aspects are:

- *Site Walkover*: The engineer will carry out a detailed observation of the customer's on-site processes, identifying areas of high consumption or wastage and cost-saving opportunities.

- *Data Analysis*: The engineer will undertake a detailed site data analysis, including efficiency correlation analysis and analysis of production, water, and effluent volumes and, where appropriate, concentrations.
- *Mass Balance*: The engineer will create a site-wide balance, exposing unidentified losses and the costs associated with each component and identifying the key cost drivers for the site.
- *Cost–Benefit Analysis*: The engineer will identify opportunities for cost savings, organizing payback periods and priorities by profitability. Welsh Water will also compare any production and water consumption data available with industry benchmarks in order to make clear and bespoke recommendations on how to improve water efficiency (Welsh Water n.d.).

Source: Welsh Water, "Process Audit," https://www.dwrcymru.com/en/Business/Our-Services/Process-Audit.aspx.

Agricultural Water Efficiency

Agricultural irrigation is one of the most substantial uses of water in most locations around the world. Therefore, on-farm water management interventions such as improved irrigation can lead to significant water consumption reductions (Huang et al. 2020). A variety of water efficiency practices are available to reduce excess water use.

On-Farm Irrigation Audits

On-farm irrigation audits are an effective method of accounting for all water usage and identifying opportunities to improve water use efficiency. They should be thought of as the first step towards agricultural water users increasing their water efficiency. They can include the measurement of water entering a farm from surface water or groundwater, the inventory and calculation of on-farm water use, the calculation of water-related costs, and the identification of potential water efficiency measures. The information obtained from such audits forms the basis for implementing measures to increase the efficiency of current farming practices and for deciding which best management practices to implement (Texas Water Development Board 2013).

Improving Water Efficiency in Irrigation

Climatic conditions, soil type and structure, plant type, and irrigation techniques are some of the main factors that influence the efficiency and effectiveness of irrigation practices. The efficiency of water irrigation practices can be improved through crop type, irrigation scheduling, irrigation method, soil enhancement measures, and water source. More particularly:

- *Crop Water Needs*: Crops differ in terms of their daily water needs and the duration of their total growing period (crops with high daily needs and a long growing season require more water). As such, irrigation water needs can be reduced by selecting crop varieties with lower water demand that still provide sufficient added value.

- *Irrigation Scheduling*: Irrigation scheduling helps remove or reduce instances where too little or too much water is applied to crops. The timing and amount of water applied in irrigation is based on the water content in the crop root zone, the amount of water consumed by the crop since it was last irrigated, and the crop development stage.
- *Irrigation Methods*: Three main irrigation methods can be selected to use available water effectively:
  - *Surface (or Gravity) Irrigation*: This involves the application of water by gravity flow to the surface of the field. There are different forms of surface irrigation, including basin irrigation, where the whole field is flooded with water, and furrow or border irrigation, where water is fed into small channels or strips of land.
  - *Sprinkler Irrigation*: Sprinkler irrigation systems imitate natural rainfall. Water is pumped through pipes and then sprayed on to the crops through rotating sprinkler heads. These systems are more efficient than surface irrigation, but they are more costly to install and operate because of the need for pressurized water. There are two main types: conventional sprinkler systems spray the water into the air, losing considerable amounts to evaporation, while low-energy precision application delivers water to the crops from drop tubes that extend from the sprinkler's arm, saving both water and energy compared to conventional systems.
  - *Drip Irrigation*: Drip irrigation delivers water using pressurized pipes and droppers that run close to the plants, which can be placed either on the soil surface or below ground. This method is highly efficient because only the immediate root zone of each plant is wetted. The system also allows precise application of water-soluble fertilizers and other agricultural chemicals.
- *Soil Enhancement Measures*: Several soil enhancement measures can be considered to improve the efficiency of irrigation practices:
  - *Proper Field Leveling*: This allows the water to travel at an optimum speed, assisting in the uniform distribution of water and reducing runoff.
  - *Furrow Dyking*: This involves the capture of irrigation or precipitation water in small earthen dams within furrows.
  - *Crop Residue Management and Conservation Tillage*: Management of the amount, orientation, and distribution of crop and plant residues on the soil surface can lead to water savings. Such practices improve the ability of the soil to hold moisture, reduce runoff from the field, and reduce surface evaporation.
- *Alternative Water Sources*: Further efficiency gains can be achieved by utilizing alternative sources of water for irrigation, including:
  - *Rainwater Harvesting*: Impermeable surfaces can be created to reduce infiltration of rainfall into the soil. The water can then be diverted to tanks, underground aquifers, or dedicated surface ponds, from which it can later be extracted and used for irrigation.
  - *Utilizing Treated Wastewater*: Domestic wastewater can be treated to meet strict health and environmental guidelines, allowing safe use in irrigation (Arab Forum for Environment and Development 2014; FAO n.d.).

Case 5.9 The Australian Government's On-Farm Irrigation Efficiency Program

The Australian government is working to improve the efficiency and productivity of on-farm irrigation water use and management. The On-Farm Water Efficiency Program, which is part of the Sustainable Rural Water Use and Infrastructure Program, is assisting irrigators in the southern connected system of the Murray–Darling Basin in modernizing their on-farm irrigation infrastructure while returning saved water to the environment. Water savings generated from the program will help secure a sustainable future for irrigation communities. More than 1500 subprojects have received government funding of around AUD$500 million to improve the efficiency of on-farm irrigation systems and on-farm water use through a range of measures, such as:

- Installing new or upgrading existing infrastructure equipment or technology, including automated water management systems and sensing equipment to improve irrigation efficiency.
- Improving irrigated area layout or design to improve on-farm irrigation efficiency (e.g., laser grading and decommissioning of old irrigation infrastructure).
- Upgrading, or converting to, surface or subsurface drip systems and overhead spray systems such as a lateral move or center pivots.
- Installing ancillary equipment necessary for new or upgraded irrigation systems to function (e.g., computer equipment or pumping equipment) (Department of Agriculture 2020).

Source: Department of Agriculture, "On-Farm Irrigation Efficiency Program," https://www.agriculture.gov.au/water/mdb/programs/basin-wide/ofiep

References

ACT Government. (n.d.). "Water Resources." https://www.actsmart.act.gov.au/what-can-i-do/schools/actsmart-schools/water/resources (accessed October 9, 2020).

Amahmid, O., Y. El Guamri, M. Yazidi, B. Razoki, K.K. Rassou, Y. Rakibi, et al. (2019). "Water Education in School Curricula: Impact on Children Knowledge, Attitudes and Behaviours Towards Water Use." *International Research in Geographical and Environmental Education* 28, no. 3: 178–193.

Arab Forum for Environment and Development. (2014). "Water Efficiency Handbook: Identifying Opportunities to Increase Water Use Efficiency in Industry, Buildings, and Agriculture in the Arab Countries." https://www.academia.edu/38680056/Water_Efficiency_Handbook_Identifying_opportunities_to_increase_water_use_efficiency_in_industry_buildings_and_agriculture_in_the_Arab_Countries_A_PRODUCT_OF_AFEDS_ARAB_GREEN_ECONOMY_INITIATIVE_AGEI (accessed October 9, 2020).

Australian Government Water Rating. (2017). "Choosing Water Efficient Products." https://www.waterrating.gov.au/choose (accessed October 9, 2020).

Brears, R.C. (2016). *Urban Water Security*. Hoboken, NJ: John Wiley & Sons.

Brears, R.C. (2018). *Natural Resource Management and the Circular Economy*. Cham: Springer.

Brears, R.C. (2020). *Developing the Circular Water Economy*. London: Palgrave Macmillan.
CIWEM. (2016). "Water Efficiency: Helping Customers to Use Less Water in Their Homes." https://www.ciwem.org/assets/pdf/Policy/Reports/Water-Efficiency.compressed.pdf (accessed October 9, 2020).
Department of Agriculture. (2020). "On-Farm Irrigation Efficiency Program." https://www.agriculture.gov.au/water/mdb/programs/basin-wide/ofiep (accessed October 9, 2020).
El-Nwsany, R.I., I. Maarouf, and W. Abd el-Aal. (2019). "Water Management as a Vital Factor for a Sustainable School." *Alexandria Engineering Journal* 58, no. 1: 303–313.
Engineering Economics. (n.d.). "Los Angeles Existing Buildings Energy and Water Efficiency Ordinance." http://www.eeiengineers.com/los-angeles-existing-buildings-energy-and-water-efficiency-ordinance/ (accessed October 9, 2020).
FAO. (n.d.). "Conservation Agriculture." http://www.fao.org/conservation-agriculture/en/ (accessed October 9, 2020).
Houston Public Works. (n.d.). "Customer Account Services." https://www.houstonwaterbills.houstontx.gov/ProdDP/Default/Default (accessed October 9, 2020).
Hove, C. and O.O. Osunkunle. (2019). "Participatory Water Conservation Education on Social Media in Amathole District Local Municipalities, Eastern Cape, South Africa." *Information Development*: 0266666919835905.
Hoy, L. and S. Stelli. (2016). "Water Conservation Education as a Tool to Empower Water Users to Reduce Water Use." *Water Science & Technology* 16, no. 1: 202–207.
Huang, G., A.Y. Hoekstra, M.S. Krol, J. Jägermeyr, A. Galindo, C. Yu, and R. Wang. (2020). "Water-Saving Agriculture Can Deliver Deep Water Cuts for China." *Resources, Conservation and Recycling* 154: 104578.
ITNews. (2016). "Victoria Trials Satellite Scans to Detect Water Leaks from Space." https://www.itnews.com.au/news/victoria-trials-satellite-scans-to-detect-water-leaks-from-space-416960 (accessed October 9, 2020).
Liu, B. and Y. Ping. (2012). "Water Saving Retrofitting and Its Comprehensive Evaluation of Existing Residential Buildings." *Energy Procedia* 14: 1780–1785.
PUB. (2015). "Water Efficiency Manager Course." https://www.pub.gov.sg/Documents/WEMC2015.pdf (accessed October 9, 2020).
Samir, N., R. Kansoh, W. Elbarki, and A. Fleifle. (2017). "Pressure Control for Minimizing Leakage in Water Distribution Systems." *Alexandria Engineering Journal* 56, no. 4: 601–612.
San Diego County Water Authority. (2016). "Water Authority Unveils Resources to Advance Region's Outdoor Water-Use Efficiency." https://www.sdcwa.org/water-authority-unveils-resources-advance-regions-outdoor-water-use-efficiency (accessed October 9, 2020).
Seelen, L.M.S., G. Flaim, E. Jennings, and L.N. De Senerpont Domis. (2019). "Saving Water for the Future: Public Awareness of Water Usage and Water Quality." *Journal of Environmental Management* 242: 246–257.
Tarhule, A. (2017). "The Future of Water: Prospects and Challenges for Water Management in the 21st Century." In *Competition for Water Resources*, edited by J.R. Ziolkowska and J.M. Peterson. Amsterdam: Elsevier, pp. 442–454.
Texas Water Development Board. (2013). "Best Management Practices for Agricultural Water Users." http://www.twdb.texas.gov/conservation/BMPs/Ag/doc/AgMiniGuide.pdf (accessed October 9, 2020).

UNIDO. (2011). "UNIDO Green Industry: Policies for Supporting Green Industry." https://www.unido.org/sites/default/files/2011-05/web_policies_green_industry_0.pdf (accessed October 9, 2020).

US EPA. (2017). "Water Efficiency Management Guide Landscaping and Irrigation" https://www.epa.gov/sites/production/files/2017-12/documents/ws-commercialbuildings-waterscore-irrigation-landscape-guide.pdf (accessed October 9, 2020).

WBDG. (2016). "Water Conservation." https://www.wbdg.org/resources/water-conservation (accessed October 9, 2020).

Welsh Water. (n.d.). "Process Audit." https://www.dwrcymru.com/en/Business/Our-Services/Process-Audit.aspx (accessed October 9, 2020).

Xu, Q., R. Liu, Q. Chen, and R. Li. (2014). "Review on Water Leakage Control in Distribution Networks and the Associated Environmental Benefits." *Journal of Environmental Sciences* 26, no. 5: 955–961.

6

Water Reuse and Water Recycling

Introduction

Water reuse is defined as the reuse of water within a single process or use of harvested water for another purpose, without treatment. In contrast, water recycling is the use of harvested water for another purpose, after treatment (State of Victoria 2009). Water reuse and water recycling provide a range of benefits to communities, including saving on the use of drinking water, reducing the amount of stormwater and treated wastewater that is discharged to the environment, protecting waterways and bays from over-abstraction and pollution, and providing an opportunity to create a greener and more livable landscape, especially in periods of climate change and drought (DELWP 2019). This chapter will discuss a range of water reuse and water recycling technologies and initiatives that have been implemented around the world, including rainwater and stormwater harvesting and potable and nonpotable reuse systems, along with a variety of alternative water source technologies.

Rainwater Harvesting

Rainwater harvesting refers to the collection of precipitation running off roof surfaces into cisterns or other storage vessels (Exall and Vassos 2012). Rainwater harvesting systems are particularly valuable in arid regions as they can reduce demand on already limited water supplies (Kumar 2019). Meanwhile, regions previously reliant on groundwater systems are also moving to rainwater harvesting systems due to pollution of the groundwater (Ammar et al. 2016; Quigley et al. 2016). Regarding the use of harvested rainwater, though it can be treated to a potable water standard and used for drinking and food preparation, it is commonly reserved for nonpotable uses, including toilet/urinal flushing, laundry, irrigation, fire suppression, and building cooling (Kumar 2019). Some of the key features of a rainwater harvesting system are listed in Table 6.1 (EERE n.d. b).

Regional Water Security, First Edition. Robert C. Brears.

Table 6.1 Key features of rooftop rainwater harvesting systems.

Feature	Description
1) Collection system	Roof surface and gutters to capture the rainwater and send it to the storage system
2) Inlet filter	Screen filter to catch large debris
3) First flush diverter	Diverter to remove debris not captured by the inlet filter from the initial stream of rainwater
4) Storage tank	Storage tank composed of food-grade polyester resin material approved by environmental regulators to help reduce bacterial growth. Typically green in color
5) Overflow	Drainage spout to allow for overflow if the storage tank gets full
6) Controls	Control system to monitor water level and filtration system
7) Treatment system	Filtration and disinfection system to treat water to potable or nonpotable standards
8) Pump	Pump to move water through the system to where it will be used
9) Backflow prevention	Backflow preventer to ensure that under negative pressure, water cannot flow back through the system into the makeup water system
10) Flowmeter	Flowmeter (with data logger) to measure water production
11) Power supply	Either a conventional power source or an alternative source such as a standalone or grid-tied solar system
12) Water level indicator	Gauge to monitor the water level in the storage tank

Source: Modified from Office of Energy Efficiency & Renewable Energy, "Water-Efficient Technology Opportunity: Rainwater Harvesting Systems," https://www.energy.gov/eere/femp/water-efficient-technology-opportunity-rainwater-harvesting-systems

Case 6.1 Austin Water's Rainwater Harvesting Rebate

Austin Water's Rainwater Harvesting Rebate helps customers buy equipment for capturing rainwater. The rebate amount is up to $5000 per site:

- Nonpressurized (no pump): $0.50 per gallon up to half of the equipment cost.
- Pressurized (has a pump): $1.00 per gallon up to half of the equipment cost.
- Customers can apply every 12 months for system expansions until they reach $5000.

Eligibility requirements include the following:

- Intention to keep a system of 500 gallons or less in service for 5 years and a system of greater than 500 gallons in service for 10 years.
- Agreement to any necessary pre- and post-installation inspections by Austin Water.
- Willingness to let Austin Water reference the project in any outreach materials (Austin Water 2019).

Source: Modified from Austin Water, "Rainwater Harvesting Rebate Guidelines and Application," (2019), http://www.austintexas.gov/sites/default/files/files/Water/Conservation/Rebates_and_Programs/2019-10-23_Rainwater_Harvesting_Rebate_Guidelines_and_Application.pdf

Stormwater Harvesting

Stormwater harvesting involves collecting, storing, and treating stormwater from urban areas for reuse in a variety of areas, including irrigation of public parks, gardens, sports fields, and golf courses. Stormwater harvesting systems can be large or small and typically consist of:

- An extraction point, where stormwater is captured or diverted from a drain, creek, or pond.
- A network of pipes to transport stormwater from the connection point to the storage site.
- A dam or storage tank where stormwater can be temporarily collected for treatment and use.
- A treatment system that produces recycled water suitable and safe for its permitted use.
- A network of pipes for distribution of recycled water.
- A system to manage byproducts produced in the stormwater harvesting facility (NEPC 2009; Minnesota Pollution Control Agency 2017).

The benefits of stormwater harvesting include:

- Reducing the demand for drinking water by replacing it with water produced from stormwater harvesting.
- Reducing stress on urban streams and rivers by capturing some of the pollutants and nutrients that would otherwise enter waterways from stormwater flows.
- Enabling users to access an alternative source of water for non-drinking water use (Sydney Water 2013; Zhang et al. 2020).

Case 6.2 Fitzroy Gardens Stormwater Harvesting System, Melbourne, Australia

The Fitzroy Gardens Stormwater Harvesting System is the largest in the City of Melbourne, providing 30 million liters of water every year and helping to keep the heritage garden healthy in a changing climate. The system captures, treats, and stores stormwater for irrigation of the park. Specifically, the former depot site at Fitzroy Gardens is a natural low point for the surrounding 67 hectare catchment. Here, stormwater runoff is captured and diverted to the system from an underground drainage network. The treatment process begins with a gross pollutant trap that removes large pollutants (litter, leaves, etc.). The water then flows to a sedimentation chamber, where suspended particles of pollution (fine sands, oils, etc.) are removed. Next to the chamber is a primary storage tank that can store 4 million liters of partially treated water. From there, the water is pumped to the surface, where a biofiltration bed naturally removes nitrogen and phosphorous. Up to 1 million liters of tested stormwater are stored in a secondary tank and used for irrigation. Any excess is returned to the stormwater drains. Finally, before the water is pumped to the Fitzroy Gardens irrigation network, it is passed over ultraviolet light tubes to kill any remaining bacteria. It is estimated that the system will replace 59% of the drinking water used for irrigation. It is part of a broader project on the former depot in Fitzroy Gardens, which includes new public open space, new depot facilities, and a visitor center. In addition to saving drinking water, the system provides a reliable alternative water source for irrigation, reduces pollutants such as nitrogen, sediment, phosphorus, and heavy metals from entering the Yarra River and Port Phillip Bay, and improves the city's ability to manage the impacts of climate change, such as heatwaves and droughts (City of Melbourne n.d.).

Source: Modified from City of Melbourne, "Fitzroy Gardens Stormwater Harvesting System," http://urbanwater.melbourne.vic.gov.au/projects/water-capture-and-reuse/fitzroy-gardens-stormwater-harvesting-project/

Water Recycling

Recycled water can be used for a range of purposes, including in urban environments (households, golf courses, recreational parks), industry (washing and cooling in power stations and mills), and agriculture (horticulture, forestry, pasture, flowers, viticulture), as well as in firefighting, groundwater recharge, irrigation of municipal landscapes, "dual-pipe" urban uses, and maintenance of environmental flows and wetlands. Water recycling provides many benefits, including:

- Reduction of nutrient and contaminant loads in oceans and rivers.
- Increase in the amount of potable water available for domestic uses, by replacing it in irrigation of crops and amenity horticulture.
- Reduction of stress on groundwater and rivers through the provision of alternative water supplies (Climate-ADAPT 2015; Voulvoulis 2018).

Case 6.3 Singapore's Public Utilities Board's Industrial Water Solutions Project Unit

Singapore's Public Utilities Board's (PUB) NEWater is ultra-clean, high-grade reclaimed water that is used mainly for industrial and air-conditioning purposes at wafer fabrication plants, industrial estates, and commercial buildings. The largest user of NEWater is wafer fabrication plants, due to their need for water of a higher quality than that used for drinking. NEWater is delivered to industrial customers via a dedicated pipe network. Currently, there are five NEWater plants, supplying up to 40% of Singapore's water needs. By 2060, NEWater is projected to meet up to 55% of the city's water demand. To reduce pressure on the NEWater system, PUB's Industrial Water Solutions Project Unit is providing funding and technical support to facilitate the development of approved projects in collaboration with industrial water users in the areas of water recycling and reuse, development of alternative sources of water such as groundwater and seawater, and minimization of water losses through cooling processes. The Industrial Water Solutions Project Unit provides funding through the Industrial Water Solutions Demonstration Fund to support high-impact and innovative projects for the treatment and reclamation of freshwater from industry for process reuse. Companies with a monthly water consumption of more than 10 000 cubic meters are invited to submit proposals for projects that yield at least a 5% reduction in water consumption through reuse. On successful completion, the project's technology may lead to full-scale implementation and could potentially be adopted in similar installations within the industrial sector (PUB n.d.a,b).

Source: PUB, "Newater," https://www.pub.gov.sg/watersupply/fournationaltaps/newater; "Funding," https://www.pub.gov.sg/research/industrialwatersolutions/funding.

Potable Reuse

Water recycling for potable reuse involves the planned use of treated municipal wastewater for augmenting drinking water supplies. There are two types of potable reuse systems:

- *Indirect Potable Reuse (IPR)*: IPR systems are where communities use surface water or groundwater containing treated wastewater as a drinking water source.

- *Direct Potable Reuse (DPR)*: DPR systems introduce purified water directly back into an existing water supply system (US EPA 2018).

Specifically, IPR introduces purified water into an environmental buffer (e.g., a groundwater aquifer or a surface water reservoir, lake, or river), and the blended water is then introduced into a water supply system, while DPR introduces purified water directly into an existing water supply system (AWWA 2016). Generally, because DPR does not require an environmental buffer, the conveyance and blending of the purified water with other potable sources are less expensive than with IPR. As such, in many cases, DPR is – or could be – more cost-effective than IPR. Also, in situations where environmental buffers are unavailable, DPR is the only potable reuse option (Lahnsteiner et al. 2017).

Potable Reuse Treatment

Public health protection is at the forefront of potable reuse, with a wide range of established treatment options and processes combining to ensure that potable reuse produces high-quality water that protects against adverse health effects from contaminant exposure. First, all wastewater is treated before discharge using one or more of the following steps:

- *Preliminary Treatment*: Removal of suspended and floating particles that may cause operational or maintenance problems in subsequent treatment processes.
- *Primary Treatment*: Removal of a portion of the suspended solids and organic matter.
- *Secondary Treatment*: Removal of most of the suspended solids and organic matter.
- *Tertiary Treatment*: Removal of targeted dissolved solids and finer suspended materials.

The treated wastewater then becomes a source of supply for potable reuse, which uses a combination of advanced processes in a second point of treatment. A summary of advanced treatment options is listed in Table 6.2 (AWWA 2016). An advanced treatment plant for potable reuse will include many of these processes in combination to not only achieve

Table 6.2 Advanced treatment options for potable reuse.

Option	Description
Biological treatment	The use of bacteria and other microorganisms to consume and remove organic materials and nutrients from water. Examples include biologically active media filters and anaerobic denitrifying filters (for the removal of nitrate and nitrite)
Membrane filtration	Processes that use a membrane barrier with microscopic pores to remove suspended particles and pathogenic microorganisms (e.g., microfiltration, ultrafiltration)
Membrane desalination	Processes that use a nonporous, semipermeable membrane barrier to remove salts, pharmaceuticals, and other dissolved contaminants (e.g., reverse osmosis (RO), nanofiltration)
Ozone	A powerful oxidant used for disinfection or the breaking down of organic contaminants, including pharmaceuticals
Advanced oxidation	Processes used to achieve significant pathogen disinfection and break down organic contaminants, including pharmaceuticals (e.g., ozone or ultraviolet light in combination with hydrogen peroxide)

Source: Modified from AWWA, "Potable Reuse 101: An Innovative and Sustainable Water Supply Solution".

water quality targets but also provide resilient, abundant, and robust treatment in a multiple-barrier approach to contaminant removal.

In addition to wastewater and advanced potable reuse treatment, IPR also uses an environmental buffer as a third point of treatment. This may be a groundwater aquifer, a surface water reservoir, or a natural or constructed wetland, and it is applied either before or after the advanced treatment. While DPR does not have an environmental barrier, it does have an engineered storage buffer. If the DPR system blends the purified water with the raw water supplies prior to their entering a drinking water plant, the engineered storage buffer occurs between the advanced treatment and drinking water treatment plants. If it is "pipe-to-pipe," the engineered storage buffer occurs between the advanced treatment plant and potable water distribution system (AWWA 2016).

Case 6.4 Indirect Potable Reuse System in Wichita Falls, Texas

In 2018, the City of Wichita Falls brought into service its IPR project, discharging water to Lake Arrowhead via a 17 mile-long pipeline. Effluent water to the lake undergoes advanced tertiary treatment utilizing Cloth Media Filtration. The IPR project can bring up to 16 million gallons of treated water a day to Lake Arrowhead, the city's primary source of potable water, making Wichita Falls more drought-tolerant by extending potable supplies and fully utilizing the return flows from its Resource Recovery Facility (City of Wichita Falls 2018, n.d.).

Source: City of Wichita Falls, "Wichita Falls Resource Recovery Facility," http://www.wichitafallstx.gov/691/Resource-Recovery ; "2018 Drinking Water Quality Report," (2018), http://www.wichitafallstx.gov/DocumentCenter/View/30928/2018-Drinking-Water-Report-for-Web.

Case 6.5 Future Water Supplies in Flagship, Arizona

Since January 1, 2018, Arizona water utilities have been able to apply to the Arizona Department of Environmental Quality for a permit for an advanced reclaimed water treatment facility for DPR that follows interim criteria specified in the Arizona reuse regulations. The City of Flagstaff is currently evaluating future water supplies, including the offering of Class A+ reclaimed water through DPR. It is planned that after reclaimed water is treated at a wastewater treatment facility, it will be treated a second time at an Advanced Water Treatment Facility. Here, advanced treatment options could include biological treatment, membrane filtration, membrane desalination, ozone, and advanced oxidation. The purified water would then be held in a storage tank for final water quality testing before being ready for distribution to the Flagship community and the Lake Mary Water Treatment Plant (City of Flagstaff n.d.; Water Reuse n.d.).

Source: WaterReuse, "State Policy and Regulations," https://watereuse.org/advocacy/state-policy-and-regulations/; City of Flagstaff, "Direct Potable Reuse," https://www.flagstaff.az.gov/3850/Direct-Potable-Reuse.

Onsite Nonpotable Reuse Systems

Onsite nonpotable reuse systems can help communities reuse water for non-drinking water purposes. Onsite nonpotable water systems capture and treat water sources generated from within or surrounding a building, such as:

- *Roof Runoff*: Precipitation from rain or snowmelt events that is collected directly from a roof surface not subject to frequent public access.
- *Stormwater*: Precipitation runoff from rain or snowmelt events that flows over land or impervious surfaces (e.g., streets, car parks, rooftops). Includes runoff from roofs that have frequent public access.
- *Domestic Wastewater or Blackwater*: Wastewater originating from toilets or kitchen sources (e.g., kitchen sinks, dishwashers).
- *Graywater*: Wastewater collected from non-blackwater sources (e.g., bathroom sinks, showers, bathtubs, clothes washers, laundry sinks).

The treated water is then reused onsite or locally for non-drinking purposes. The most common indoor use is toilet and urinal flushing, which can be up to 25% of the total water demand in a residential building and up to 75% of the total water demand in a commercial building. Other potential uses of the water are irrigation, cooling/heating applications, process water, and clothes washers, all of which can increase nonpotable water demand by up to 50% for residential buildings and up to 95% for commercial buildings (US Water Alliance 2017; US EPA n.d.).

Case 6.6 New York City's On-Site Water Reuse Systems Grant

The 2014 New York City Construction Code regulates two types of onsite water reuse systems:

- *Wastewater Reuse Systems (Blackwater, Graywater, Rainwater)*: These may be used for toilet and urinal flushing, cooling tower makeup, laundry, subsurface or drip irrigation systems, watering plants with a hose, the washing of footpaths, streets, or buildings, and other approved uses that are located in the same lot as the water reuse system.
- *Rainwater Reuse Systems (Rainwater, Cooling Tower Condensate)*: These may be used solely for cooling tower makeup and subsurface or drip irrigation.

New York City's Department of Environmental Protection (DEP) provides grant assistance with the construction of nonpotable water reuse systems at both the building and the district scale. The grant program's overall goal is to achieve a total water demand reduction of 1 million gallons per day. It facilitates two separate categories of projects:

1) Building-scale projects can receive up to $250 000 in reimbursement for a system designed to save at least 32 000 gallons per day.
2) District-scale projects can receive up to $500 000 in reimbursement for a system designed to save at least 94 000 gallons per day (district-scale projects involve two or more parcels, such as a housing development, where the project reduces demand in the shared distribution system).

Buildings with successful onsite water reuse systems are also eligible for the Comprehensive Water Reuse Program (CWRP), which provides a 25% water and wastewater fee discount to DEP customers who install water reuse systems that reduce a building's water consumption by at least 25% (NYC DEP 2018).

Source: NYC DEP, "2018-2019 Nyc on-Site Water Reuse Systems Grant Application Guide," (2018), https://www1.nyc.gov/assets/dep/downloads/pdf/water/drinking-water/onsite-water-reuse-presentation.pdf

Environmental Flows

Treated wastewater can be used to restore and enhance natural habitats such as wetlands and marshes or to maintain flows of small water bodies, which may preserve or restore biodiversity. The creation of these types of habitats, for environmental and recreational purposes, can also be supported by water reuse. Along coastlines, water reuse to restore and enhance lagoons and wetlands may be a suitable alternative to the discharge of effluent into the sea (European Commission 2016). Some of the benefits of environmental flows are listed in Table 6.3 (North Central Catchment Management Authority 2017).

Table 6.3 Benefits of environmental flows.

Beneficiary	Examples
Community	Improves water quality for stock and domestic needs and downstream irrigators and other users
	Provides attractive and enjoyable destinations for tourists, families, and industries
	Increases investment opportunity in environmentally based tourism
	Attracts recreational fishers
	Allows increased productivity of farming lands in associated wetlands
Environment	Supports breeding of insect-eating waterbirds
	Prevents regular stagnation of water
	Moves carbon (the product of decomposition of material buried in or lying on floodplains) between river floodplains and wetlands, maintaining the health of rivers
	Improves the health of riverbank vegetation, which stabilizes banks and slows erosion
	Stimulates native fish to move on to floodplains and breed
	Provides flows of sufficient depth and duration for waterbirds to build nests, breed, and raise chicks
	Replenishes aquifers and dilutes salty water left in wetlands following evaporation
	Stimulates invertebrate eggs to hatch and plant seeds to germinate

Source: Modified from North Central Catchment Management Authority, "Environmental Flows," (2017), http://www.nccma.vic.gov.au/sites/default/files/publications/nccma-65245_-_environmental_flows_q_.pdf

Case 6.7 Utilization of Reclaimed Water for Restoration of Waterways in Tokyo

The Tokyo Metropolitan Government recommends the use of recycled water for miscellaneous uses in buildings. Furthermore, because of the abundance and stability of recycled water, it is used for the restoration of waterways. Examples include:

- *Restoration of Three Rivers in the Jonan Area (Shibuya and Furukawa River, Meguro River, and Nomikawa River)*: Following advanced treatment in Ochiai Water Reclamation Center, reclaimed water is fed into the three rivers, where river flow has decreased and the stream environment deteriorated, contributing to an increase of water volume and recuperation of the water environment.
- *Restoration of Waterways in Tama Area (Nobidome Waterway, Tamagawa and Senkawa Channels)*: Treated wastewater from the Water Reclamation Center on the Tama River undergoes sand filtration and ozonation before being fed into Nibidome Waterway, which has long experienced stagnated water flow, as well as the Tamagawa and Senkawa Channels. The streams now have healthy flows and provide a place of recreation and relaxation for residents (Bureau of Sewerage n.d.).

Source: Bureau of Sewerage Tokyo Metropolitan Government, "Utilization of Reclaimed Water," http://www.gesui.metro.tokyo.jp/english/aboutus/env-guide/06/.

Other Alternative Water Sources

Other alternative water sources include the following.

Captured Condensate

Water condenses on the cooling coils of mechanical equipment (e.g., dedicated outdoor air units) when humid air contacts these cool surfaces. A large amount of condensate can form on coils in areas with hot, humid summers. Water that collects on cooling units must be drained to prevent damage to the equipment or building. Usually, the condensate is collected in a central location and discharged into a sewer drain. In a condensate capturing system, it is instead directed to a central storage tank or basin and then distributed for reuse. The water yield of such a system can vary depending on its design and purpose, ranging from 15–50 liters/day for small portable units to 200 000 liters/day for larger agricultural-scale designs. The different scales mean these systems can be used for cooling tower makeup and for toilet and urinal flushing, water feature makeup, and landscape irrigation (Lawrence 2010; Khalil et al. 2016; EERE n.d.a).

Case 6.8 Condensate Capture Potential Map

The potential for condensate capture from air-conditioning systems across the United States was studied by faculty and staff at the University of Georgia. The researchers developed a method to estimate how much condensate could be collected annually, based on a location's annual average dew point, cooling degree days, and rainfall. The method was applied to 47 cities across the country, with the results mapped to determine patterns in the amount of condensate that could be captured in specific regions. To evaluate this further, the Federal Energy Management Program applied the same methodology to 64 additional sites. All in all, 111 locales were placed on the US map, allowing for the condensate potential to be characterized across the country. The map is color-coded with regions ranging from highest to lowest potential. Users can locate their site and determine the potential for condensate capture based on the color of their region. Potentials are categorized as follows:

- *Highest*: Areas with high cooling loads during most of the year and high humidity. Coastal areas also have high potential.
- *High*: Areas with high cooling loads and high summer humidity.
- *Medium-High*: Areas with relatively hot and humid summers.
- *Medium*: Areas with humid summers.
- *Low and Lowest*: Areas with low humidity or relatively low cooling demand (EERE n.d.c).

Source: U.S. Department of Energy, "Condensate Capture Potential Map," https://www.energy.gov/eere/femp/condensate-capture-potential-map

Atmospheric Water Generation Technology

Atmospheric water generators (AWGs) produce potable water from surrounding air, potentially increasing water availability during shortages and contamination events as well as other events that interrupt drinking water services such as storms and public water infrastructure failures. In the past, bottled water and water tankers were delivered to affected populations in the case of such events. However, this solution was hampered by limited water capacities and the relatively expensive continuous delivery costs. Sustainable onsite water supply technology is thus more practical. AWGs range in size from domestic units that produce up to 20 liters of water per day to commercial units that produce 1000–10 000 liters per day. They operate using vapor-compression refrigeration technology to condensate water from ambient air by cooling it below its dew-point temperature. Makeup water for cooling towers can also be an ideal use of AWGs, as the water produced has extremely low dissolved mineral content (Bagheri 2018; Tu et al. 2018; US EPA 2019; EERE n.d. a).

Case 6.9 World Vision South Africa's Mobile Water Generation Project

World Vision South Africa launched an innovative mobile water generation project in the drought-stricken communities in the Eastern Cape region. With funding of $330 000 from the Bill Ford Better World Challenge, the Ford Motor Company Fund, and Ford Research and Advanced Engineering, the project utilizes a Watergen GEN-350 AWG that can produce up to 900 liters of water per day. The AWG is powered by a diesel generator, ensuring the system is completely mobile and can be operated in any location. The system includes multi-medium air filtration to ensure the water it generates is of the highest standard, regardless of air quality (Ford 2019).

Source: Based on Ford, "Ford and World Vision Introduce Innovative Water Generating Project for Drought-Stricken Communities," https://media.ford.com/content/fordmedia/fna/us/en/news/2019/06/14/ford-and-world-vision-introduce-innovative-water-generating-proj.html

Water Purification System Discharge Water

Water purification systems, such as RO, remove impurities (e.g., minerals, calcium, chloride, sodium, and chlorine) from a water supply for processes that require ultra-pure water. The discharged water containing impurities is usually directed to a sewer drain line. The ratio of purified water to the total water supply is called the recovery rate. A typical recovery rate for a water purification system is between 50 and 75% (equating to 50 and 25% of the total water supplied being rejected, respectively). As such, discharge water can be recovered and reused. Because the discharge water from an RO system is only slightly less pure than the source water entering the filter, it is suitable for irrigation and other nonpotable reuse applications, including toilet and urinal flushing and cooling tower makeup. If the discharge water is used for landscape irrigation, the landscape plants should have a high tolerance for salinity (Alliance for Water Efficiency n.d.; EERE n.d. a).

Foundation Water

Buildings that have water collecting around their foundations and basements because of groundwater or drainage from stormwater runoff use sump pumps to prevent flooding. This water normally goes directly into a storm or sewer system, but it can be recovered and reused for a variety of purposes, including toilet and urinal flushing, cooling tower makeup, irrigation, and car washing (Alliance for Water Efficiency n.d.).

Blowdown Water

Blowdown water is water that is drained from cooling equipment and boilers to remove mineral buildup developed during evaporation cooling or steam production. As water evaporates, its mineral concentrations increase, requiring removal from the system. Systems that require blowdown include cooling towers, evaporative condensers, evaporative coolers, evaporative cooled air-conditioners, and central boilers. Blowdown water is usually discharged directly into a sewer. However, it can be recovered and reused for other purposes, including irrigation. Nonetheless, with its high mineral content, blowdown water may need to be diluted with another source of water before it can be used (Alliance for Water Efficiency n.d.).

Desalinated Water

Desalination is a process that reduces the dissolved salt content of saline water to a usable level. All desalination processes involve three liquid streams of the saline feedwater (brackish water or seawater), low-salinity product water, and very saline concentrate (brine or reject water). There are two types of membrane processes used for desalination: RO and electrodialysis (ED). In the RO process, water from a pressurized saline solution is separated from the dissolved salts by flowing water through a water-permeable membrane. The permeate – the liquid flowing through the membrane – is encouraged to flow by the pressure differential created between the pressurized feed water and the product water, which is at near atmospheric pressure. The remaining feedwater is brine. ED is an electrochemical membrane process that reduces salt concentration from saltwater by a direct current. It occurs when feedwater flows into an ion-exchange chamber that consists of cation- and anion-exchange membranes positioned alternately between two electrodes: the cathode and the anode. When an electrical force is applied to the electrodes, the cations (positively charged ions) travel toward the cathode and the anions (negatively charged ions) travel toward the anode. The cations pass through the cation-exchange membrane and are held back by the anion-exchange membrane. The anions pass through the anion-exchange membrane and are held back by the cation-exchange membrane. At the end of the process, concentrated cations and anions flow to the brine residue stream (UNEP 1998; Sharbat 2016).

Case 6.10 Masdar's Renewable Energy Seawater Desalination Program

In 2013, Masdar launched a renewable energy desalination pilot program to facilitate research and development in energy-efficient, cost-competitive desalination technologies, including those powered by renewable energy. Commercial partners developed and operated a next-generation pilot seawater desalination plant in Ghantoot, Abu Dhabi to test a range of approaches to increasing operational efficiency, including solar-powered RO. The renewable energy seawater desalination program demonstrates two different technologies:

- *Type One Advanced Technologies*: Defined as technologies that are based on commercially provide systems.
- *Type Two Innovative Desalination Technologies*: Defined as technologies that are based on novel concepts of seawater desalination.

The program consists of two stages:

- *Stage One (2013–2017) (Piloting Phase)*: Demonstration of energy-efficient desalination systems at a small scale. The pilot plants will operate continuously for at least 18 months to demonstrate the reliability performance of the developed technologies.
- *Stage Two (2018–2020) (Implementation)*: Implementation of the developed energy-efficient desalination technologies on a large scale in the United Arab Emirates and the broader region, through seawater desalination plants entirely powered by renewable energy. A commercial-scale facility was intended to be operational by 2020 (Masdar 2017).

Source: Modified from Masdar, "Ghantoot Desalination Pilot Plant," https://masdar.ae/en/masdar-clean-energy/projects/ghantoot-desalination-pilot-plant

Fog Harvesting

Water can be collected from fog: a mass of water vapor condensed into small water droplets at or just above the earth's surface. The small droplets precipitate when they encounter objects. Fog has the potential to provide an alternative source of freshwater, as it can be harvested using simple, low-cost collection systems known as fog collectors. These work best in coastal areas, where the water can be harvested as the fog moves inland, driven by the wind. However, they can also be used in mountainous areas. Fog collectors may be simple, flat, rectangular nets of nylon supported by a post at either end and arranged perpendicular to the direction of the prevailing wind, or they can be complex structures made up of a series of joined collection panels. As water collects on the net, the droplets join to form larger drops that fall – under the influence of gravity – into a trough or gutter at the bottom of the panel. From there, the water is conveyed to a storage system by gravity. Fog capture from cooling tower plumes in thermal power plants is possible, with a study finding that recovery of about 40% of the drift loss (amounting to a saving of nearly 10.5 cubic meters of water per hour from a 500 MW unit) could be achieved (Ghosh et al. 2015). Unlike natural fog harvesting schemes, where the fog-laden flow is predominantly horizontal, fog flow stream in a cooling tower rises against gravity (UNEP 1998).

Case 6.11 Women-Led Fog Harvesting in Morocco

Dar Si Hmad, a women-led nongovernmental organization in Morocco, has designed and installed the world's most extensive operational fog water harvesting system. The Dar Si Hmad project provides potable water to more than 400 people in five villages, most of them women and children. The project includes:

- 600 square meters of nets.
- Seven reservoirs with 593 cubic meters of storage capacity.
- Six solar panels.
- More than 10 000 meters of piping.
- Prepaid water meters connected to 52 homes in five villages, serving more than 400 rural Berber residents.

Dar Si Hmad's female team leaders have conducted over 20 capacity-building workshops with rural Berber women to promote literacy and income-generating projects. In 2017, the project was upgraded with CloudFisher net technology, which mimics the way a spider's web naturally collects dew droplets from mist. This net can harvest up to 36 000 liters of water a day for 800 residents. With more water collected, it enables more time for women to work and children to study (UNFCCC n.d.).

Source: UNFCCC, "Women-Led Fog Harvesting for a Resilient, Sustainable Ecosystem | Morocco," https://unfccc.int/climate-action/momentum-for-change/women-for-results/women-led-fog-harvesting-for-a-resilient--sustainable-ecosystem.

References

Alliance for Water Efficiency. (n.d.). "Reverse Osmosis Discharge Water." https://www.allianceforwaterefficiency.org/resources/topic/reverse-osmosis-discharge-water (accessed October 9, 2020).

Ammar, A., M. Riksen, M. Ouessar, and C. Ritsema. (2016). "Identification of Suitable Sites for Rainwater Harvesting Structures in Arid and Semi-Arid Regions: A Review." *International Soil and Water Conservation Research* 4, no. 2: 108–120.

Austin Water. (2019). "Rainwater Harvesting Rebate Guidelines and Application." http://www.austintexas.gov/sites/default/files/files/Water/Conservation/Rebates_and_Programs/2019-10-23_Rainwater_Harvesting_Rebate_Guidelines_and_Application.pdf (accessed October 9, 2020).

AWWA. (2016). "Potable Reuse 101: An Innovative and Sustainable Water Supply Solution." https://www.awwa.org/Portals/0/AWWA/ETS/Resources/PotableReuse101.pdf?ver=2018-12-12-182505-710 (accessed October 9, 2020).

Bagheri, F. (2018). "Performance Investigation of Atmospheric Water Harvesting Systems." *Water Resources and Industry* 20: 23–28.

Bureau of Sewerage. (n.d.). "Utilization of Reclaimed Water." Tokyo Metropolitan Government. http://www.gesui.metro.tokyo.jp/english/aboutus/env-guide/06/ (accessed October 9, 2020).

City of Flagstaff. (n.d.). "Direct Potable Reuse." https://www.flagstaff.az.gov/3850/Direct-Potable-Reuse (accessed October 9, 2020).

City of Melbourne. (n.d.). "Fitzroy Gardens Stormwater Harvesting System." http://urbanwater.melbourne.vic.gov.au/projects/water-capture-and-reuse/fitzroy-gardens-stormwater-harvesting-project/ (accessed October 9, 2020).

City of Wichita Falls. (2018). "2018 Drinking Water Quality Report." http://www.wichitafallstx.gov/DocumentCenter/View/30928/2018-Drinking-Water-Report-for-Web (accessed October 9, 2020).

City of Wichita Falls. (n.d.). "Wichita Falls Resource Recovery Facility." http://www.wichitafallstx.gov/691/Resource-Recovery (accessed October 9, 2020).

Climate-ADAPT. (2015). "Water Recycling." https://climate-adapt.eea.europa.eu/metadata/adaptation-options/water-recycling (accessed October 9, 2020).

DELWP. (2019). "Using Alternative Water Sources." https://www.water.vic.gov.au/liveable/using-alternative-water-sources (accessed October 9, 2020).

EERE. (n.d.-a a). "Best Management Practice #14: Alternative Water Sources." https://www.energy.gov/eere/femp/best-management-practice-14-alternative-water-sources (accessed October 9, 2020).

EERE. (n.d.-b b). "Water-Efficient Technology Opportunity: Rainwater Harvesting Systems." https://www.energy.gov/eere/femp/water-efficient-technology-opportunity-rainwater-harvesting-systems (accessed October 9, 2020).

EERE. (n.d.-c c). "Condensate Capture Potential Map." https://www.energy.gov/eere/femp/condensate-capture-potential-map (accessed October 9, 2020).

European Commission. (2016). "Common Implementation Strategy for the Water Framework Directive and the Floods Directive: Guidelines on Integrating Water Reuse into Water Planning and Management in the Context of the WFD." https://ec.europa.eu/environment/water/pdf/Guidelines_on_water_reuse.pdf (accessed October 9, 2020).

Exall, K. and T.D. Vassos. (2012). "Integrated Urban Water Management: Water Use and Reuse." In *Metropolitan Sustainability*, edited by F. Zeman. Cambridge: Woodhead, pp. 319–349.

Ford. (2019). "Ford and World Vision Introduce Innovative Water Generating Project for Drought-Stricken Communities." https://media.ford.com/content/fordmedia/fna/us/en/news/2019/06/14/ford-and-world-vision-introduce-innovative-water-generating-proj.html (accessed October 9, 2020).

Ghosh, R., T.K. Ray, and R. Ganguly. (2015). "Cooling Tower Fog Harvesting in Power Plants – A Pilot Study." *Energy* 89: 1018–1028.

Khalil, B., J. Adamowski, A. Shabbir, C. Jang, M. Rojas, K. Reilly, and B. Ozga-Zielinski. (2016). "A Review: Dew Water Collection from Radiative Passive Collectors to Recent Developments of Active Collectors." *Sustainable Water Resources Management* 2, no. 1: 71–86.

Kumar, A. (2019). "Water Conservation Technologies." In *Hydraulic Rubber Dam*, edited by S. Thomas, A.V. Rane, Abitha V.K., K. Kanny, and A. Dutta. Norwich: William Andrew, pp. 11–24.

Lahnsteiner, J., P. van Rensburg, and J. Esterhuizen. (2017). "Direct Potable Reuse – A Feasible Water Management Option." *Journal of Water Reuse and Desalination* 8, no. 1: 14–28.

Lawrence, T., J. Perry, and P. Dempsey. (2010). "Capturing Condensate by Retrofitting AHUs." *ASHRAE Journal* 52, no. 1.

Masdar. (2017). "Ghantoot Desalination Pilot Plant." https://masdar.ae/en/masdar-clean-energy/projects/ghantoot-desalination-pilot-plant (accessed October 9, 2020).

Minnesota Pollution Control Agency. (2017). "Overview for Stormwater and Rainwater Harvest and Use/Reuse." https://stormwater.pca.state.mn.us/index.php/Overview_for_stormwater_and_rainwater_harvest_and_use/reuse (accessed October 9, 2020).

NEPC. (2009). "Australian Guidelines for Water Recycling: Stormwater Harvesting and Reuse." Natural Resource Management Ministerial Council, the Environment Protection and Heritage Council, and the National Health and Medical Research Council. http://www.nepc.gov.au/system/files/resources/5fe5174a-bdec-a194-79ad-86586fd19601/files/wq-agwr-gl-stormwater-harvesting-and-reuse-final-200907.pdf (accessed October 9, 2020).

North Central Catchment Management Authority. (2017). "Environmental Flows." http://www.nccma.vic.gov.au/sites/default/files/publications/nccma-65245_-_environmental_flows_q_.pdf (accessed October 9, 2020).

NYC DEP. (2018). "2018–2019 NYC On-Site Water Reuse Systems Grant Application Guide." https://www1.nyc.gov/site/dep/index.page (accessed October 9, 2020).

PUB. (n.d. a). "Funding." https://www.pub.gov.sg/research/industrialwatersolutions/funding (accessed October 9, 2020a).

PUB. (n.d. b). "NEWater." https://www.pub.gov.sg/watersupply/fournationaltaps/newater (accessed October 9, 2020b).

Quigley, N., S.G. Beavis, and I. White. (2016). "Rainwater Harvesting Augmentation of Domestic Water Supply in Honiara, Solomon Islands." *Australasian Journal of Water Resources* 20, no. 1: 65–77.

Sharbat, A. (2016). "Development of Photovoltaic Electrodialysis Desalination System." Report 191. https://www.usbr.gov/research/dwpr/reportpdfs/report191.pdf (accessed October 9, 2020).

State of Victoria. (2009). "Guidelines for Water Reuse and Recycling in Victorian Health Care Facilities: Non-Drinking Applications." https://www2.health.vic.gov.au/Api/downloadmedia/%7B949656D2-00DA-486E-B450-84C75D71A0BF%7D (accessed October 9, 2020).

Sydney Water. (2013). "Stormwater Harvesting." https://www.sydneywater.com.au/web/groups/publicwebcontent/documents/document/zgrf/mdq3/~edisp/dd_047752.pdf (accessed October 9, 2020).

Tu, Y., R. Wang, Y. Zhang, and J. Wang. (2018). "Progress and Expectation of Atmospheric Water Harvesting." *Joule* 2, no. 8: 1452–1475.

UNEP. (1998). "Source Book of Alternative Technologies for Freshwater Augmentation in Latin America and the Caribbean." https://www.oas.org/dsd/publications/Unit/oea59e/begin.htm#Contents (accessed October 9, 2020).

UNFCCC. (n.d.). "Women-Led Fog Harvesting for a Resilient, Sustainable Ecosystem/Morocco." https://unfccc.int/climate-action/momentum-for-change/women-for-results/women-led-fog-harvesting-for-a-resilient--sustainable-ecosystem (accessed October 9, 2020).

US EPA. (2018). "Mainstreaming Potable Water Reuse in the United States: Strategies for Leveling the Playing Field." https://www.epa.gov/sites/production/files/2018-04/documents/mainstreaming_potable_water_reuse_april_2018_final_for_web.pdf (accessed October 9, 2020).

US EPA. (2019). "Atmospheric Water Generation Technology." https://www.epa.gov/sites/production/files/2019-11/documents/awg_technical_brief_final_05nov19.pdf (accessed October 9, 2020).

US EPA. (n.d.). "Onsite Non-Potable Water Reuse Research." https://www.epa.gov/water-research/onsite-non-potable-water-reuse-research (accessed October 9, 2020).

US Water Alliance. (2017). "A Guidebook for Developing and Implementing Regulations for Onsite Non-Potable Water Systems." http://uswateralliance.org/sites/uswateralliance.org/files/publications/NBRC%20GUIDEBOOK%20FOR%20DEVELOPING%20ONWS%20REGULATIONS.pdf (accessed October 9, 2020).

Voulvoulis, N. (2018). "Water Reuse from a Circular Economy Perspective and Potential Risks from an Unregulated Approach." *Current Opinion in Environmental Science & Health* 2: 32–45.

WaterReuse. (n.d.). "State Policy and Regulations." https://watereuse.org/advocacy/state-policy-and-regulations/ (accessed October 9, 2020)..

Zhang, K., P.M. Bach, J. Mathios, C.B.S. Dotto, and A. Deletic. (2020). "Quantifying the Benefits of Stormwater Harvesting for Pollution Mitigation." *Water Research* 171: 115395.

Part II

Green Infrastructure

7

Green Buildings and Green Streets

Introduction

Green infrastructure at the building and street levels comprises a wide range of water features, from green roofs to underground storage systems, aimed at the management of both water quantity and water quality. It also offers a range of economic, environmental, and social co-benefits. This chapter will first provide an overview of green and blue roofs, before discussing rain barrels and cisterns, rain gardens, bioswales, and permeable pavement. Finally, it will look at the application of green infrastructure at the street scale.

Green Roofs

Green roofs, also referred to as vegetated roofs, eco-roofs, and roof gardens, are more than just plants placed on a rooftop. They are highly engineered contiguous systems of plantings, drainage layers, and watertight membranes. Their systems are designed to protect the structural integrity of the building below, while providing runoff volume reduction and runoff peak rate attenuation, along with multiple environmental, economic, and social co-benefits (Table 7.1) (NYC Parks 2013; Berardi et al. 2014; Li and Yeung 2014; Berardi 2016; Vinnova 2016; Francis and Jensen 2017; DOEE 2018; Langemeyer et al. 2019; NOAA 2019).

Types of Green Roofs

There are three main types of green roofs:

- *Extensive*: The growing medium is 3–4 inches thick, the additional roof load is between 15 and 30 pounds per square foot, there are fewer varieties of plants, less irrigation is required, and the roof is drought-resistant and requires low maintenance.
- *Intensive*: The growing medium is 8–12 inches thick, the additional roof load is as much as 40–150 pounds per square foot, there are greater varieties of plants (including shrubs and small trees), consistent irrigation is required during summer months, routine maintenance is required, and the roof provides a highly designed usable green space.

Regional Water Security, First Edition. Robert C. Brears.

Table 7.1 Benefits of green roofs.

Type	Benefit	Description
Economic	Reduced lifecycle costs	Green roofs can last up to three times as long as conventional ones
	Reduced waste and decreased need for landfill	The extended lifetime of a green roof reduces long-term construction waste and the cost of construction
	Increased property values	As an added amenity, green roofs attract higher rents and higher tenant retention
	Energy savings	Green roofs reduce the energy required for heating and cooling. In the winter, they insulate the upper floors of the building, and in the summer, they cool the building through shading and the process of evapotranspiration by plants
	Sound insulation	The substrate reduces noise pollution, adding to the desirability of the building
	Credits for stormwater charges	Green roofs can provide credits for stormwater fees, saving money on regulatory fees
	Adjacent solar arrays	Green roofs cool the air around solar arrays, increasing their efficiency
Environmental	Reduced urban heat island effect	Green roofs stay significantly cooler than conventional ones, reducing ambient air temperature
	Reduced stormwater runoff	In summer months, green roofs can retain 70–100% of stormwater; in winter, they can retain 40–50%. This reduces the volume and velocity of stormwater, reducing the potential for localized flooding and for erosion and sedimentation of local waterways
	Improved air quality	Green roof vegetation metabolizes airborne particles, such as smog and sulfur dioxide
	Carbon sequestration	Green roofs take carbon out of the carbon cycle
	Creation of wildlife habitat	Green roofs provide urban green infrastructure, which helps enhance species biodiversity
Social	Education opportunities	Green roofs provide areas for instruction in ecology, biology, and math
	Space for food production	Green roofs create opportunities for urban agriculture, potentially increasing food security within urban areas
	Aesthetic appeal	The vegetation and natural beauty of green roofs offer a respite from the concrete hardscape of urban areas
	Usable green space	Green roofs can provide green space throughout urban areas, where open space is limited
	Job creation and economic security	The establishment of a green roofing industry creates new jobs in manufacturing, construction, design, installation, maintenance, and horticulture

Source: DOEE, "Green Roof Toolkit," (2018), https://doee.dc.gov/sites/default/files/dc/sites/ddoe/publication/attachments/2018%201.5-Toolkit%20DRAFT.pdf; Vinnova, "Swedish Guidelines for Green Roofs: Short English Summary", (2016); NOAA, "Green Infrastructure Options to Reduce Flooding," (2019), https://coast.noaa.gov/digitalcoast/training/gi-reduce-flooding.html; NYC Parks, "Nyc Parks Green Roof: A Living Laboratory for Innovative Green Roof Design," (2013), https://www.nycgovparks.org/pagefiles/53/Citywide-Services-Green-Roof_2.pdf; Lotte Fjendbo Møller Francis and Marina Bergen Jensen, "Benefits of Green Roofs: A Systematic Review of the Evidence for Three Ecosystem Services," Urban Forestry & Urban Greening 28(2017); Umberto Berardi, "The Outdoor Microclimate Benefits and Energy Saving Resulting from Green Roofs Retrofits," Energy and Buildings 121(2016); W. C. Li and K. K. A. Yeung, "A Comprehensive Study of Green Roof Performance from Environmental Perspective," International Journal of Sustainable Built Environment 3, no. 1 (2014); Johannes Langemeyer et al., "Creating Urban Green Infrastructure Where It Is Needed – a Spatial Ecosystem Service-Based Decision Analysis of Green Roofs in Barcelona," Science of The Total Environment (2019); Umberto Berardi, AmirHosein GhaffarianHoseini, and Ali GhaffarianHoseini, "State-of-the-Art Analysis of the Environmental Benefits of Green Roofs," Applied Energy 115(2014).

- *Semi-intensive*: The growing medium is 4–8 inches thick, the additional roof load is 25–50 pounds per square foot, there are more plant varieties (including wildflowers and drought-tolerant herbaceous perennials), and periodic maintenance is required (City of Toronto 2013; Li and Yeung 2014; Philadelphia Water Department 2018; Shafique et al. 2018).

Fundamental Design Issues to Consider

Stormwater management effectiveness can be improved by taking measures to enhance evapotranspiration, including:

- Introducing low-transmissivity drainage layers.
- Lengthening seepage pathways to drains.
- Introducing rainfall interception layers.
- Selecting plants with dense root habitats.
- Selecting a medium with high water-holding capacity.

Meanwhile, coordinating the green roof design with the design of building heating, ventilation, and air conditioning (HVAC) systems can help maximize operational cost savings (Philadelphia Water Department 2018).

Green Roof Components

The typical components of green roofs include:

- *Inlet Control Component*: Green roofs that receive direct (1 : 1) rainfall do not have inlet controls. For those that receive runoff from a directly connected impervious area (DCIA), inlet control systems may convey and control the flow of stormwater from the contributing catchment area. Distribution piping can be used to evenly disperse run-on water throughout the roof.
- *Storage Area Component*: Green roof storage areas temporarily hold stormwater before it is either used by plants through evapotranspiration or released downstream. Storage areas are typically composed of:
  - *Growing Medium*: This supports plant growth and provides for storage of stormwater within voids. The storage capacity is a function of medium depth, surface area, and total void space.
  - *Filter or Separation Fabric, or Geotextile*: This prevents migration of soil into the underlying drainage area of the green roof.
  - *Drainage Layer*: This may incorporate measures to intercept and retain percolated rainfall as it moves through the green roof storage area, examples of which include membranes with depressions to hold water.
  - *Moisture Interception Layers/Root Barriers*: These are impermeable liners that protect the underlying roof deck from moisture and plant root intrusion.
  - *Underlying Roofing System*: This typically consists of a structural deck, its supporting structures, and a traditional overlying waterproofing system.
- *Vegetated Component*: Green roof plant material is designed to take up as much as possible of the water that falls on the roof during a storm event. It mitigates wind and water

erosion, transpires captured moisture back into the atmosphere, and provides evaporative cooling. Plant material also collects dust and creates oxygen.

- *Outlet Control Component*: Outlet controls regulate the release of stormwater from a stormwater management practice. They must be appropriately selected and sized for the storage area component. Controls may include risers, edge drains, scuppers, gutters, and impervious liners.
- *Inspection and Maintenance Access Component*: The provision of safe and easy inspection and maintenance access to all major components of a green roof is critical to ensuring long-term performance. Inspection and maintenance access components may include permanent or temporary safety monitoring systems, guardrail and safety net systems, warning systems, and long-term leak-detection systems (Philadelphia Water Department 2018; Shafique et al. 2018).

Blue Roofs

Blue roofs, also known as controlled-flow roof drain systems, are detention stormwater management systems that provide temporary storage and slow release of rainwater on a rooftop. On a blue roof, water is temporarily detained on the roof surface using rooftop check dams or roof drainage restrictions. In combined sewage areas where infiltration is not feasible, blue roofs can be used to meet water quality release-rate and pollutant-reducing requirements. They can be installed on many types of roofs, including terraces, high-rise building roofs, and low-podium or at-grade on-structure installations. They can be applied as a stormwater retrofit on existing buildings with flat, mildly sloped, or terraced roofs after confirmation of adequate structural loading capacity, waterproofing, and protection of rooftop utilities (Brears 2018; Philadelphia Water Department 2018).

Blue Roof Components

The typical components of blue roofs include:

- *Inlet Control Component*: Blue roofs that receive direct (1 : 1) rainfall only do not have inlet controls. For those that receive runoff from adjacent roof DCIA, inlet control systems convey and control the flow of stormwater from the contributing catchment area.
- *Storage Area Component*: Blue roof storage areas temporarily hold stormwater until it can either evaporate or be released downstream at a controlled rate. The area dedicated to storage depends on the chosen blue roof system type:
  - Storage in roof drain restrictor systems is determined by the roof slope and geometry. The bulk volume occupied by all building mechanical systems, roof furniture, and so forth must also be factored into the storage volume calculations.
  - Storage in check dam systems is determined by the roof slope and associated area dedicated to ponding behind the dams. The bulk volume occupied by all building mechanical systems, roof furniture, and so forth must also be factored into the storage volume calculations.

- *Outlet Control Component*: Outlet controls within a blue roof system can provide a range of functions, including meeting drain downtime requirements, controlling the rate of discharge from the system and limiting water surface elevations during various storm events, and bypassing of flows from large storm events.
- *Inspection and Maintenance Access Component*: Depending on roof height and slope, blue roof inspection and maintenance access components may include permanent or temporary safety monitoring systems, guardrail and safety net systems, warning line systems, personal fall arrest systems, or long-term leak-detection systems for locating and managing leaks (Philadelphia Water Department 2018).

Case 7.1 Green, Multifunctional Roofs in Rotterdam

Rotterdam already has 360 000 square meters of green roof, and since January 1, 2020, homeowners, companies, associations, and foundations have been able to apply for a subsidy to construct more. The amount available is as follows, up to a combined maximum of €50 000:

- €500 per added cubic meter of water storage.
- €5 per added square meter of coupled area.
- €10 per added square meter of green area.

Rotterdam encourages various types and functions of eco-roofs (multifunctional roofs), including green roofs with solar panels, blue roofs that function as water gardens with a water buffer during extreme rainfall, and green roofs that provide a space for working, conferencing, gardening, and so on. The following functions and colors are defined:

- Green roofs provide greening, contribute to air quality, and offer opportunities to grow fruit or vegetables.
- Blue roofs can collect water, delay excess rainwater, relieve sewers, and reduce flooding.
- Yellow roofs generate sustainable energy from the sun or wind.
- Red roofs include all kinds of social functions, such as terraces, playgrounds, and bars.
- Orange roofs are used for mobility, such as roof bridge connections.
- Purple roofs are residential roofs.
- Gray roofs have technical functions, such as air handling units and chimneys (Gemeente Rotterdam 2020).

Source: Based on Gemeente Rotterdam, "Groene Daken," https://www.rotterdam.nl/wonen-leven/groene-daken/

Rain Barrels and Cisterns

Rain barrels and cisterns are low-cost systems that reduce runoff volume and, during small storm events, delay and reduce peak runoff flow rates. Rain barrels are usually used for individual residences, while cisterns have both residential and commercial applications (WERF n.d.).

Rain Barrels

Rain barrels are typically small scale (25–100 gallons) and located at the downspout of a gutter system. The total volume of storage available from rain barrels represents roof runoff from relatively small rainfall events, typically less than 1 inch of rain over the surface. This is a small volume for a single rain barrel. Still, the cumulative effects of rain barrels installed across a watershed include stormwater volume reduction and water quality treatment, since the first inch of runoff contains the dirtiest water (UNL Water 2019).

Cisterns

Cisterns have a higher storage capacity than rain barrels. They may be located either above- or belowground in lawns, recreational areas, parking lots, or other impervious areas, or within buildings when space constraints exist. Due to their size and storage capacity, these systems (often large polyethylene drums) collect runoff from areas more extensive than residential rooftops, such as commercial parking lots. The storage capacity of a cistern is a function of the catchment area, the depth of rainfall required to fill the system, and the volume and timing of water use. A general rule of thumb in sizing a system is that 1 inch of rainfall on a 1000 square foot roof will yield approximately 600 gallons of runoff (Philadelphia Water Department 2018).

Case 7.2 San Francisco Public Utilities Commission's Free Rain Barrels and Discounted Cisterns

The San Francisco Public Utilities Commission (SFPUC) has partnered with the Urban Farmer Store to provide residents, businesses, and schools with free rain barrels (participating customers need only pay the tax of $8.50). Also, customers can receive a $350 discount on larger cisterns. Participants in the SFPUC Rainwater Harvesting Program will receive either:

- a free 50 gallon rain barrel (limit two per customer), plus a hose bib, overflow hose, and connecting valve to join multiple barrels; or
- a $350 discount off the purchase of a cistern, ranging from 205 to 750 gallons (limit one per customer) (SFPUC 2019).

Source: Modified from SFPUC, "Rainwater Harvesting," https://www.sfwater.org/index.aspx?page=178.

Rain Gardens

Rain gardens are depression gardens designed and located to receive water runoff from a roof, driveway, or lawn. They slow stormwater flow, help remove pollutants, recharge freshwater bodies, and look attractive. Rain gardens work by ponding stormwater in a planted area, to be filtered through the soil mix and plant roots before flowing into surrounding ground, pipes, drains, and streams. Typically, a slotted or perforated pipe underdrainage system within a granular drainage layer is included at the bottom of a rain garden if it is

Table 7.2 Components of rain gardens.

Component	Description
Rain garden media mix	Filters pollutants
Ponding area	Holds stormwater runoff until it seeps through the media mix and into the underdrain system
Plants	Help filter pollutants and look attractive. Native plants are usually better suited to extreme/dry conditions, such as ponding for 24 hours or no rain for a month
Entry and overflow system	Stormwater entry choice impacts on maintenance regime and costs. Ideally, a grassed filter strip or grassed swale is included before entry, with the first-stage filtration removing larger particles. Sedimentation bays should be used in high silt load-risk areas. An overflow bypass for excess flows is utilized when the rain garden ponding area is full. At times, an internal overflow sump is required
Gravel mulch layer	Prevents weeds and helps prevent soil drying out
Filter sand and transition layer	An additional filter layer which helps retain media within the rain garden
Drainage layer	Collects and transports filtered water toward the stormwater collection system. Very permeable soils may not require a drainage layer or piped collection system
Edge support	Protects the carriageway subgrade from saturation and failure during media replacement
Underdrain system	Collects filtered water
Rain garden base	Assists in partial soakage

connected to a street-level stormwater pipeline. Rain gardens are usually most suited for retrofit into existing developed areas where urban intensification or redevelopment is taking place (Christchurch City Council 2016).

Rain Garden Components

The critical components of a rain garden are given in Table 7.2.

Benefits of Rain Gardens

Rain gardens provide multiple benefits, including:

- Removing pollutants (sediment, fertilizers, pesticides, and vehicle fluids and metals) from stormwater.
- Increasing water filtration and recharging groundwater supplies.
- Enhancing the beauty of properties, neighborhoods, and businesses through provision of landscaped areas.
- Providing habitat for birds, butterflies, and beneficial insects.
- Reducing flooding and drainage problems in properties and communities.

- Sustaining waterway flows during dry periods.
- Reducing the flow intensity of waterways during storm events.
- Improving health through stress reduction and the provision of places to exercise.
- Providing space for people to enjoy nature.
- Facilitating environmental education (Bray et al. 2012; University of Arkansas 2012; Brears 2016; Christchurch City Council 2016; Siwiec et al. 2018).

Case 7.3 Rain Gardens in New York City

The New York City Department of Environmental Protection (DEP) is building rain gardens and other types of green infrastructure to manage stormwater and improve water quality in the city's waterways. The construction of the rain gardens follows a design and construction process:

- *Selection*: Potential locations are marked with green spray paint (not all locations that receive spray paint will receive a rain garden, as selection can take several months). A drilling company takes samples of the soil to ensure it can absorb stormwater (only those locations that can effectively do so are considered). Engineers work with utility companies to avoid conflicts with existing service lines.
- *Construction*: Before construction, DEP reaches out to surrounding property owners. During construction, the sidewalk and curb are removed, and the material underneath is excavated to a depth of about 5 feet. The excavated area is backfilled with stone and engineered soil, which allows for infiltration. The sidewalk and curb are replaced, with one or two curb cuts in the new curb to allow water to flow in or out. The contractor installs plants and trees and cares for them so that they will thrive in their new setting.
- *Maintenance*: Before taking over maintenance, the city inspects each completed rain garden to ensure it collects stormwater properly. The city is responsible for rain garden maintenance. Maintenance crews remove litter, sediment, and weeds from each installation regularly. Crews also replant the greenery as needed (NYC DEP 2020).

Source: NYC DEP, "Rain Gardens in Nyc," (2020), https://www1.nyc.gov/assets/dep/downloads/pdf/water/stormwater/green-infrastructure/nyc-rain-gardens.pdf.

Bioswales

Bioswales are shallow, open channels with gently sloping sides that can be incorporated into the landscape to direct flow, slow runoff, and promote pollutant removal. Vegetation planted within them (e.g., native plants and grasses) helps slow the flow of water and filter out pollutants, while permeable soils encourage filtration. Specifically, bioswales filter stormwater and improve overall water quality via the following processes:

1) Transit of stormwater through surface vegetation.
2) Percolation of stormwater through prescribed filter media, which provides treatment through fine filtration, extended detention treatment, and some biological uptake.
3) Disconnection of impervious areas from downstream waterways.
4) Protection of natural wetland systems from frequent storm events through reduction of storm flow velocities compared to pipe systems.

Bioswale Design

Bioswales are typically linear in shape, with a sloped bottom, and are dry most of the time. They are expected to contain standing water during storms and to be emptied within 24 hours. They are often used as alternatives to, or enhancements of, traditional stormwater piping. They are usually incorporated into parking lots and road medians and located parallel to roadways in order to infiltrate and treat a portion of the stormwater volume (University of Florida 2008; County of San Diego 2014; City of Edmonton 2016).

Bioswale performance is affected by soil type, groundwater table, size of area serviced, imperviousness of contributing watershed, and system dimensions and slope. Bioswales should be used to serve areas of less than 10 acres, with slopes no greater than 5%. A common rule of thumb is that the total surface area of a bioswale should be 1% of the area from which it is receiving stormwater. Large areas should be divided and treated using multiple bioswales. Vegetated swales should not be installed in areas with high water tables where groundwater reaches the bottom of the swale. Additional considerations include:

- Bioswale function and treatment are improved when applied in areas with well-drained soils.
- Poorly drained sites require an underdrain system.
- Slopes >5% require multiple ditch blocks or weirs perpendicular to the flow to facilitate storage volume and extend the time of infiltration.
- Where phosphorous control is important, soils used should have a low phosphorous content and high phosphorous sorption capacity.
- Where nitrogen control is important, mulch and organic matter incorporated into the soil should have a high carbon-to-nitrogen ratio.
- A high-flow bypass (outlet or control structure) should be included to convey high flows safely (University of Florida 2008).

Case 7.4 Vancouver's Green Street with Bioswales

The City of Vancouver has a target of capturing and treating 90% of its average annual rainfall using green infrastructure tools and design guidelines on public and private property. To achieve this, a range of green infrastructure projects have been implemented around the city. One of these, the 63rd Avenue and Yukon Street green infrastructure project, was identified in a community plan as having the opportunity to increase access to green space, enhance rainwater management, and improve neighborhood amenities. With construction completed in 2018, the project includes 102 square meters of dedicated bioretention systems and manages runoff from more than 1170 square meters of adjacent impermeable area. It includes a variety of public and green features, including sidewalks, seating areas, bike racks, a drinking water fountain, a rain garden, and a bioswale. The rain garden and bioswale collectively clean and manage 90% of the average annual runoff from surrounding roads (2200 cubic meters of rainwater per annum), resulting in improved water quality and reduced combined sewer overflows (CSOs) (City of Vancouver 2018).

Source: Based on City of Vancouver, "63rd Avenue and Yukon Street Green Infrastructure Plaza," https://vancouver.ca/home-property-development/63rd-avenue-and-yukon-street-boulevard-improvements.aspx.

Permeable Surfaces

There are a variety of permeable surfaces that reduce stormwater runoff flows and improve water quality.

Permeable Pavements

Permeable pavements look like standard pavements except that void spaces are created by omitting fine materials. Under the permeable pavement is a filter course comprising finer aggregate, which overlays a stone reservoir, the thickness of which depends on the stormwater storage needs and load-bearing requirement. Below the stone reservoir, a filter fabric is placed on the undisturbed soil. This prevents soil particles from entering the stone reservoir due to fluctuations in the water table or any pumping action from repeated loads. Filter fabric should also be placed along the sides or perimeter of the permeable pavement system to prevent soil from entering at these locations. Perforated pipes can be added near the top of the stone reservoir to discharge excess stormwater from significant events. Also, instead of allowing stormwater to infiltrate into the underlying soil, or where the permeability of the underlying soil is not optimal, perforated underdrain pipes can be installed to route water to an overflow facility structure (University of Kentucky 2011; Auckland Council 2016). There are numerous types of permeable pavement available.

Pervious Concrete

Pervious concrete is a mixture of cement, coarse aggregate or gravel, and water. Unlike conventional concrete, pervious concrete contains a void content of 15–35%, achieved by removing finer particles such as sand from the mixture. The empty spaces allow water to infiltrate the underlying soil instead of either pooling on the surface or being discharged as runoff. Footpaths and parking lots are ideal applications. Pervious concrete can withstand the light loads generated by pedestrian and bicycle traffic, but the loads generated in parking lots can be substantial and require consideration when determining the concrete mix and pavement thickness (while the strength of the pervious concrete can be increased by adding cement, its porosity will decrease, resulting in decreased infiltration rates).

Porous Asphalt

Porous asphalt is a standard asphalt mixture of both fine and coarse aggregate bound together by a bituminous binder, except it uses less fine aggregate than conventional asphalt. It has a similar percentage of void space to pervious concrete. Its surface is like that of conventional asphalt, but it has a rougher texture. Its compressive strength is also similar to conventional asphalt. Therefore, porous asphalt can be used for both pedestrian applications (e.g., greenways) and low-speed vehicular traffic applications (e.g., parking lots, curbside parking lanes, and residential or side streets).

Permeable Pavers

Permeable interlocking concrete pavers (PICPs), permeable interlocking clay brick pavers (PICBPs), and concrete grid pavers (CGPs) are similar in terms of installation and function, but are made from different materials:

- PICPs are solid concrete blocks that fit together to form a pattern with small aggregate-filled spaces in between them that allow stormwater to infiltrate (these spaces usually account for 5–15% of the surface area).
- PICBPs are the same as PICPs except the material is brick instead of concrete.
- CGPs include large openings due to their lattice-style configuration, accounting for 20–50% of their surface area and usually containing soil or grass, though small aggregates can be used (University of Kentucky 2011).

Case 7.5 Permeable Pavement Project in the City of Tacoma, Washington

The Oakland Neighborhood Permeable Pavement Project in the City of Tacoma is converting 18 failed residential streets into green infrastructure using permeable pavement. The project covers 20 city blocks and will fully infiltrate an 18-acre neighborhood. Currently, stormwater from these 20 blocks travels 3000 feet through the city's storm sewer pipe before discharging into the wetland. This project will reduce the contaminant loading to the wetland. It will also improve the base flow, which is significant to the health of this sensitive receiving water (City of Tacoma 2019).

Source: Based on City of Tacoma, "Oakland Neighborhood Permeable Pavement Project," https://www.cityoftacoma.org/government/city_departments/environmentalservices/surface_water/green_stormwater_infrastructure__gsi_/gsi_projects/oakland_neighborhood_permeable_pavement_project.

Green Streets

In most communities, streets make up a significant percentage of publicly owned land. Green streets use a variety of systems, including vegetated curb extensions, sidewalk planters, landscaped medians, vegetated swales, permeable pavement, and street trees, to reduce stormwater, improve water quality, enhance pedestrian safety, and beautify neighborhoods.

Residential Streets

Residential streets offer the highest potential for building green streets in new neighborhoods or retrofitting existing streets as they are usually slower, have less traffic, and are already likely to have some landscape elements. In newly constructed streets, green streets can be designed to manage significant volumes of water. At the same time, in retrofit situations, they can typically handle all the rain for small storms, with excess water from large ones overflowing into existing storm sewer systems. Various stormwater management features can be installed in new or retrofitted green streets, as summarized in Table 7.3 (US EPA 2009).

Commercial Streets

Commercial streets need to accommodate a wide range of users and uses, including pedestrians, cyclists, drivers, on-street parking, lighting, trees, and so on. As such, finding enough space to collect and manage stormwater can be a challenge. It is therefore essential to find areas that can accommodate multiple purposes in open space. A variety of stormwater systems can be installed in new or retrofitted commercial streets (Table 7.3) (US EPA 2009).

Table 7.3 Green infrastructure used in residential and commercial streets.

Street Type	Green Infrastructure	Description
Residential	Stormwater curb extensions	Conventional curb extensions have been used for many years to enhance pedestrian safety and help in traffic calming. A stormwater curb extension incorporates a rain garden into which runoff flows
	Permeable paving	Permeable paving in the parking lane converts impervious surfaces to allow stormwater to absorb into the ground, which reduces the amount of runoff without any loss of parking on the street. It also gives the illusion of a narrower street, helping to calm traffic
	Vegetated swales	As water flows through the swale, it is slowed down by the interaction with plants and soils, allowing sediments and pollutants to settle out
Commercial	Stormwater planters	Stormwater flows into planters and is absorbed into the plants and topsoil. It fills up to a predetermined level, and if necessary overflows into a storm sewer system. Planters can also accommodate trees
	Stormwater curb extensions	Like those on residential streets, curb extensions can be located near corners. They manage stormwater and provide pedestrians with a more comfortable cross. They can also be located mid-street by converting parking spaces
	Permeable paving	Permeable paving can be incorporated into footpaths and car parks. It can be designed for higher-speed traffic or for where heavy vehicles are expected to be parked (e.g., loading zones and bus stops)

Source: Modified from US EPA, "Green Streets: A Conceptual Guide to Effect Green Streets Design Solutions Residential Streets, Commercial Streets, Arterial Streets, Alleys," (2009), https://nepis.epa.gov/Exe/ZyPDF.cgi/P10059Y4.PDF?Dockey=P10059Y4.PDF

Arterial Streets

Arterial streets in towns and cities frequently have vast expanses of pavement, minimal greenery, and few design features to address pedestrian needs. They can be retrofitted to reduce runoff and help protect water quality. If the adjacent landscape does not exist, it is possible to convert part of the paved surface into green space based on how much is necessary to manage traffic safely. Retrofitting these streets also makes them more "multimodal" by incorporating footpaths, on-street cycle lanes, or landscaped-separated cycle greenways. Arterial streets are suitable for vegetated swales as they typically have long, linear stretches of uninterrupted space that can be used to manage stormwater. If there is no landscape space in place, travel lanes or paved shoulders can be narrowed to create it (US EPA 2009).

Alleys

In many towns and cities, alleys represent a significant amount of impervious surface. They are sometimes prone to flooding, as they are not often connected to the sewer system. Green street techniques such as vegetated swales and permeable pavement can effectively reduce and treat runoff, alleviate flooding, and are less expensive than installing sewer connections. In particular:

- *Permeable Paving*: Alleys are usually low-speed and low-trafficked streets and therefore suitable locations for using permeable paving. The entire surface can be permeable. If heavier vehicles are likely to use the alley, or if it slopes toward the center line, then only the middle section needs to be permeable.
- *Vegetated Swales*: If an alley is sloping toward the center line, then stormwater can be accommodated by greening the edges with swales and planters. Water can also flow through pipes or covered trenches to allow vehicle access to garages and driveways (US EPA 2009).

Case 7.6 Melbourne's Green Your Laneway Program

The City of Melbourne established the Green Your Laneway program to help transform the city's laneways into leafy, green, usable spaces for everyone to enjoy. An interactive map was developed to show laneways that could go green, based on the amount of sunlight they received, their exposure to wind, and their physical characteristics. In 2017, based on the interactive map data and through a process of community voting, the city provided funding for four of these laneways: Katherine Place, Meyers Place, Guildford Lane, and Coromandel Place. These pilot lanes have since been completed, providing the community with a range of multiple benefits, as summarized in Table 7.4 (City of Melbourne 2018).

Table 7.4 Benefits of Melbourne's Green Your Laneway program.

Benefit	Example
Environmental	Diverting stormwater runoff from the lanes into the soil
	Filtering dust and pollution from the air
	Improving biodiversity levels in the central city
	Providing habitat for wildlife
	Reducing noise levels in the city
	Insulating buildings from heat and cold, reducing energy expenditure and carbon emissions
	Reducing the urban heat island effect through shading and cooling
Social	Reinvigorating laneways from waste areas to usable public spaces
	Providing more pleasant walkways and thoroughfares, encouraging people to walk and spend time outdoors
	Reducing vandalism and antisocial behavior
	Bringing nature into the city, with positive effects in reducing depression and illness
Economic	Reducing heating and cooling costs (due to improved building insulation)
	Extending the life expectancy of impervious surfaces
	Increasing property values
	Increasing usable green outdoor space

Source: Modified from City of Melbourne, "Greening Laneways," https://www.melbourne.vic.gov.au/community/greening-the-city/green-infrastructure/Pages/greening-laneways.aspx.

Source: City of Melbourne, "Greening Laneways," https://www.melbourne.vic.gov.au/community/greening-the-city/green-infrastructure/Pages/greening-laneways.aspx © City of Melbourne.

Green Parking Lots

Green parking lots use green infrastructure that closely mimics the natural water cycle to manage stormwater through rainfall retention, pollutant removal, and water infiltration. Some of the most common examples are listed in Table 7.5 (US EPA 2008; Newman et al. 2013).

Table 7.5 Green infrastructure used in green parking space.

Example	Description
Swales	Open channels or depressions with dense vegetation used to transport, decelerate, and treat runoff. In green parking lots, they are designed to help direct water into bioretention areas
Vegetated filter strips	Flat pieces of land with low slopes, designed to encourage natural sheet flow of stormwater. They are often incorporated to pretreat and remove sediment before the water enters infiltration devices (e.g., bioretention areas)
Bioretention areas (rain gardens)	Also known as rain gardens, these are typically used in parking lot islands. Unlike swales, they are well suited for parking lots in denser, urban areas with less available space
Dry detention basins	Vegetated basins with controlled outlets, designed to detain runoff for a short amount of time, partially removing pollutants before the water is discharged
Wet retention basins	Designed to capture, filter, store, and infiltrate stormwater. Have storage capacity adequate for flood volumes
Infiltration systems	Designed to capture and retain stormwater runoff, allowing water to gradually infiltrate into the ground for hours or days, depending on the design. Two conventional systems are used in green parking lots: • *Infiltration Basins*: Open depressions that cover a relatively large area. They are constructed to work in conjunction with filter strips or swales, which help direct runoff from a parking space into the basin • *Infiltration Trenches*: Shallow excavated ditches lined with filter strips and filled with stones to form a subsurface basin, where water is stored until it infiltrates into the soil

Sources: US EPA, "Green Parking Lot Resource Guide," (2008), https://nepis.epa.gov/Exe/ZyPDF.cgi/P100D97A.PDF?Dockey=P100D97A.PDF; Newman, A.P., D. Aitken, and B. Antizar-Ladislao. (2013). "Stormwater Quality Performance of a Macro-Pervious Pavement Car Park Installation Equipped with Channel Drain based Oil and Silt Retention Devices." *Water Research* 47, no. 20: 7327–7336.

Case 7.7 Incentivization of Stormwater Management in Parking Lots in Minneapolis, Minnesota

The City of Minneapolis' Stormwater Utility Fee pays for the city's stormwater system and maintenance costs and helps prevent and correct stormwater runoff problems. All properties within city limits, bar a few exceptions, pay a monthly stormwater utility fee that is based on impervious area and is charged on a per-unit basis. The impervious area includes parking lots as well as driveways, footpaths, and so on. Each Equivalent Stormwater Unit (ESU) is 1530 square feet of impervious area on a property. This area is calculated based on the property size and current use. All properties, except for single-family properties, which have their own fee structure, are billed as follows: Gross Lot Size (square feet) multiplied by Runoff Coefficient (based on land use class) divided by 1530 square feet equals number of ESUs. Stormwater fees can be reduced through the Stormwater Credits Program, which offers fee reductions to owners who use approved methods to manage stormwater runoff on their property. Fees can also be reduced through replacement of impervious areas with green space (Minneapolis Public Works n.d.).

Source: Based on Minneapolis Public Works, "Stormwater Utility Fee Faq," http://www.ci.minneapolis.mn.us/publicworks/stormwater/fee/stormwater_fee_stormwater_faq.

Underground Storage

On-site, underground storage retention/detention captures and stores stormwater collected from surrounding impervious areas. Pipes or curb cuts lead surface stormwater to subsurface vaults or systems, or to large-diameter interconnected storage pipes or chambers. Stored water is then released directly through an outlet pipe back into natural waters at rates designed to reduce peak water flows during storms. Stormwater can also be allowed to infiltrate to recharge groundwater. Underground storage is usually used in developments where land availability, shape, and land costs hamper the development of surface green infrastructure practices. On-site storage attenuates peak stormwater flows and is ideal for use under parking lots, roadways, and paved areas in commercial, industrial, and residential developments. Table 7.6 provides an overview of the variety of considerations that need to be addressed when installing an underground storage system (Lake Superior Streams n.d.). Their use has a number of advantages, including:

- Reduction of stormwater runoff flow.
- Extended storage and the slow, measured release of collected stormwater runoff.
- Suitability for high-density or urban areas with limited available space, or places where land parcels are unusually shaped or expensive.
- Rapid system installation, as it uses prefabricated modular systems.
- Durability and long life (most systems have a life span of 50 years or more).

Table 7.6 Underground storm design considerations.

Materials and Installation Considerations	Description
Site selection and planning	Check with local authorities regarding their design requirements and necessary permits for construction of underground storage
	The use of experienced engineers and suppliers is recommended
	The placement of underground storage is site-specific. During early inspections, a note should be made of site size, shape, and physical characteristics. These factors help determine the basic structure of the system and the best materials to use
	The suggested maximum area of stormwater drainage to be collected for one underground storage system is 25 acres

(Continued)

Table 7.6 (Continued)

Materials and Installation Considerations	Description
Material selection	Materials used for underground storage are influenced by: • Depth and area of excavation • Shape of space (continuous space allows for the use of concrete, while angular spaces favor pipe systems) • Depth of water table (a high, close-to-system table can cause plastic pipes to float upward)
Design	Capacity and discharge rate will depend on local government stormwater requirements. Typical design requirements include: • The system should result in a reduction of stormwater flow rate to one that mimics pre-development conditions – no net increase in the runoff for a design storm event in a newly developed area • The system should handle a particular size of storm event over a particular length of time An emergency overflow system should be designed to convey flows more significant than can be handled by the storage system or to divert water if the system becomes inoperable for any reason Enough personnel access points (manholes) should be incorporated in the design to facilitate easy maintenance
Maintenance	Periodic inspections of the inlet and outlet areas should be made, to ascertain the correct operation of the system and to clean materials trapped on grates protecting catch basins and inlet areas The primary maintenance concerns are removal of trapped floatables and removal of accumulating sediments Any structural repairs to inlet and outlet areas should be addressed promptly on an as-needed basis
Cost	Underground stormwater facilities are more expensive than surface stormwater treatment methods. However, in areas where land is expensive or in short supply, underground storage is cost-effective Cost is highly variable, depending on: • Materials used • Storage volume required • Construction and labor costs • Site location • Physical conditions at site

Source: Modified from Lake Superior Streams, "Underground Storage," http://www.lakesuperiorstreams.org/stormwater/toolkit/underground.html.

Case 7.8 Underground Water Storage in Rotterdam

The municipality of Rotterdam, the Schieland Water Board, and the Krimpenerwaard have developed an underground water storage solution (ondergrondse waterberging, OWB) that temporarily collects rainwater in heavy/prolonged showers. The OWB has a volume of 10 000 cubic meters and has been incorporated into the design of a parking garage. Its two functions, parking and collecting water, always remain separate from each other, as the water is stored in residual spaces beneath the parking garage, such as the spaces below the entrances and exits. When the city's canals and sewers can no longer handle excess rainwater, the OWB temporarily absorbs it, filling in around 30 minutes. As soon as the rain stops, the OWB is pumped into the sewer until empty. The pumps have a capacity of 900 cubic meters per hour, so this takes approximately 10 hours. Once dry, the OWB is rinsed with a vacuum flushing system, where 40 000 liters of water are released in one go, passing through the tank like a tidal wave and rinsing all the sludge away. The OWB is then ready for the next rain shower. Without this system, CSOs would occur on average 10 times a year. With it, they only occur once every 2 years. Overall, the OWB provides 12% of the water storage capacity needed for the city center (Gemeente Rotterdam n.d.).

Source: Modified from Gemeente Rotterdam, "Ondergrondse Waterberging Museumparkgarage," https://www.rotterdam.nl/wonen-leven/waterberging-museumparkgarage/

References

Auckland Council. (2016). "Permeable Pavement Construction Guide." https://www.aucklandcouncil.govt.nz/environment/looking-after-aucklands-water/stormwater/docsconstructionguides/permeable-pavement-construction-guide.pdf (accessed October 9, 2020).

Berardi, U. (2016). "The Outdoor Microclimate Benefits and Energy Saving Resulting from Green Roofs Retrofits." *Energy and Buildings* 121: 217–229.

Berardi, U, A.H. GhaffarianHoseini, and A. GhaffarianHoseini. (2014). "State-of-the-Art Analysis of the Environmental Benefits of Green Roofs." *Applied Energy* 115: 411–428.

Bray, B., D. Gedge, G. Grant, and L. Leuthvilay. (2012). "Rain Garden Guide." https://raingardens.info/wp-content/uploads/2012/07/UKRainGarden-Guide.pdf (accessed October 9, 2020).

Brears, R.C. (2016). *Urban Water Security*. Hoboken, NJ: John Wiley & Sons.

Brears, R.C. (2018). *Blue and Green Cities: The Role of Blue-Green Infrastructure in Managing Urban Water Resources*. London: Palgrave Macmillan.

Christchurch City Council. (2016). "Rain Garden Design, Construction and Maintenance Manual." https://www.ccc.govt.nz/assets/Documents/Consents-and-Licences/construction-requirements/IDS/15-1297087-Rain-garden-design-construction-and-maintenance-manual.PDF (accessed October 9, 2020).

City of Edmonton. (2016). "LID – Bioswales Factsheet." https://www.edmonton.ca/city_government/documents/LID_Bioswales_Factsheet.pdf (accessed October 9, 2020).

City of Melbourne. (2018). "Greening Laneways." https://www.melbourne.vic.gov.au/community/greening-the-city/green-infrastructure/Pages/greening-laneways.aspx (accessed October 9, 2020).

County of San Diego. (2014). "Low Impact Development Handbook: Stormwater Management Strategies." https://www.sandiegocounty.gov/content/dam/sdc/pds/docs/LID_Handbook_2014.pdf (accessed October 9, 2020).

City of Tacoma. (2019). "Oakland Neighborhood Permeable Pavement Project." https://www.cityoftacoma.org/government/city_departments/environmentalservices/surface_water/green_stormwater_infrastructure__gsi_/gsi_projects/oakland_neighborhood_permeable_pavement_project (accessed October 9, 2020).

City of Toronto. (2013). “City of Toronto Guidelines for Biodiverse Green Roofs.” https://www.toronto.ca/city-government/planning-development/official-plan-guidelines/green-roofs/biodiverse-green-roofs/ (accessed October 9, 2020).

City of Vancouver. (2018). “63rd Avenue and Yukon Street Green Infrastructure Plaza.” https://vancouver.ca/home-property-development/63rd-avenue-and-yukon-street-boulevard-improvements.aspx (accessed October 9, 2020).

DOEE. (2018). “Green Roof Toolkit.” https://doee.dc.gov/sites/default/files/dc/sites/ddoe/publication/attachments/2018%201.5-Toolkit%20DRAFT.pdf (accessed October 9, 2020).

Francis, L.F.M. and M.B. Jensen. (2017). “Benefits of Green Roofs: A Systematic Review of the Evidence for Three Ecosystem Services.” *Urban Forestry & Urban Greening* 28: 167–176.

Gemeente Rotterdam. (2020). “Groene Daken.” https://www.rotterdam.nl/wonen-leven/groene-daken/ (accessed October 9, 2020).

Gemeente Rotterdam. (n.d.). “Ondergrondse Waterberging Museumparkgarage.” https://www.rotterdam.nl/wonen-leven/waterberging-museumparkgarage/ (accessed October 9, 2020).

Lake Superior Streams. (n.d.). “Underground Storage.” http://www.lakesuperiorstreams.org/stormwater/toolkit/underground.html (accessed October 9, 2020).

Langemeyer, J., D. Wedgwood, T. McPhearson, F. Baró, A.L. Madsen, and D.N. Barton. (2019). “Creating Urban Green Infrastructure Where It Is Needed – A Spatial Ecosystem Service-Based Decision Analysis of Green Roofs in Barcelona.” *Science of The Total Environment*: 135487.

Li, W.C. and K.K.A. Yeung. (2014). “A Comprehensive Study of Green Roof Performance from Environmental Perspective.” *International Journal of Sustainable Built Environment* 3, no. 1: 127–134.

Minneapolis Public Works. (n.d.). “Stormwater Utility Fee FAQ.” http://www2.minneapolismn.gov/residents/utilities/index.htm (accessed October 9, 2020).

Newman, A.P., D. Aitken, and B. Antizar-Ladislao. (2013). “Stormwater Quality Performance of a Macro-Pervious Pavement Car Park Installation Equipped with Channel Drain based Oil and Silt Retention Devices.” *Water Research* 47, no. 20: 7327–7336.

NOAA. (2019). “Green Infrastructure Options to Reduce Flooding.” https://coast.noaa.gov/digitalcoast/training/gi-reduce-flooding.html (accessed October 9, 2020).

NYC DEP. (2020). “Rain Gardens in NYC.” https://www1.nyc.gov/assets/dep/downloads/pdf/water/stormwater/green-infrastructure/nyc-rain-gardens.pdf (accessed October 9, 2020).

NYC Parks. (2013). “NYC Parks Green Roof: A Living Laboratory for Innovative Green Roof Design.” https://www.nycgovparks.org/pagefiles/53/Citywide-Services-Green-Roof_2.pdf (accessed October 9, 2020).

Philadelphia Water Department. (2018). “Philadelphia Stormwater Management Guidance Manual, Version 3.2.” https://www.pwdplanreview.org/manual-info/guidance-manual (accessed October 9, 2020).

SFPUC. (2019). “Rainwater Harvesting.” https://www.sfwater.org/index.aspx?page=178 (accessed October 9, 2020).

Shafique, M., R. Kim, and M. Rafiq. (2018). “Green Roof Benefits, Opportunities and Challenges – A Review.” *Renewable and Sustainable Energy Reviews* 90: 757–773.

Siwiec, E., Erlandsen, A.M., and Vennemo, H. (2018). “City Greening by Rain Gardens – Costs and Benefits.” *Environmental Protection and Natural Resources* 29, no. 1: 1–5.

University of Arkansas. (2012). "Rain Gardens and Stormwater." https://www.uaex.edu/publications/pdf/FSA-9533.pdf (accessed October 9, 2020).

University of Florida. (2008). "Bioswales/Vegetated Swales." http://buildgreen.ufl.edu/Fact_sheet_Bioswales_Vegetated_Swales.pdf (accessed October 9, 2020).

University of Kentucky. (2011). "Permeable Pavement for Stormwater Management." http://www2.ca.uky.edu/agcomm/pubs/aen/aen108/aen108.pdf (accessed October 9, 2020).

UNL Water. (2019). "Rainwater Harvesting in Residential Scale Landscapes." https://water.unl.edu/article/lawns-gardens-landscapes/rainwater-harvesting-residential-scale-landscapes (accessed October 9, 2020).

US EPA. (2008). "Green Parking Lot Resource Guide." https://nepis.epa.gov/Exe/ZyPDF.cgi/P100D97A.PDF?Dockey=P100D97A.PDF (accessed October 9, 2020).

US EPA. (2009). "Green Streets: A Conceptual Guide to Effect Green Streets Design Solutions Residential Streets, Commercial Streets, Arterial Streets, Alleys." https://nacto.org/docs/usdg/2000_green_streets_epa.pdf (accessed October 9, 2020).

Vinnova. (2016). "Swedish Guidelines for Green Roofs: Short English Summary." https://docplayer.net/154497824-Swedish-guidelines-for-green-roofs-short-english-summary.html (accessed October 9, 2020).

WERF. (n.d.). "Rain Barrels and Cisterns." https://www.werf.org/liveablecommunities/toolbox/rainbarrel.htm (accessed October 9, 2020).

8

Green Parks and Urban Forests

Introduction

Parklands contain significant permeable surfaces that can easily absorb rainwater. If well designed, parks can be enhanced to create hydraulic connections to larger land areas that are mainly impervious, enabling them to filter stormwater runoff from surrounding roadways and other impervious surfaces. Meanwhile, urban forests and vegetation intercept and filter stormwater runoff, preventing flooding and improving water quality in addition to providing multiple economic, environmental, and social co-benefits (Brears 2018). This chapter will discuss first green parks and then urban forests in the context of managing excess stormwater.

Green Parks

In cities, water-capturing surfaces are usually scarce, being located mainly in private properties, in schools and campuses, alongside roadways, and within public parks. In addition to parks capturing rain where it falls, excess water from surrounding areas can be directed into them, with two main benefits: first, the park can store this water for later uses in dry periods, and second, the city reduces the volume of stormwater requiring treatment in wastewater treatment plants and the amount of pollution in its water bodies.

Methods for Managing Stormwater in Parks

There are two main methods of managing stormwater in parks: holding it for conventional treatment and using natural processes to percolate it through the soil. Detaining stormwater for conventional treatment is considered to be only a "half-green" method. It treats runoff as waste by preventing it from percolating through the soil, sending it to traditional wastewater treatment plants for physical and chemical treatment processes before discharging it into a waterway. In contrast, the "fully green" way is to view stormwater as a resource, with green infrastructure removing contaminants, diverting the water away from the sewer system, and returning it directly to the ground. Nonetheless, many cities use a

Regional Water Security, First Edition. Robert C. Brears.

mix of techniques, with some green infrastructure containing overflow drain connectors to the traditional storm sewer system in case of a heavy rainfall event (TPL 2016).

Considerations for Using Green Infrastructure

There is no set template for green infrastructure in urban parks as geography dictates that there are numerous kinds, including narrow streamside greenways, vast flat forestlands, community gardens, and sports complexes. Each park has its own history and culture, resulting in different approaches to designs for managing stormwater. When it comes to the use of green infrastructure in parks, there are three main considerations:

1) *Physical Relationship*: Could a redesign of the park reduce neighborhood flooding or the pollution of downstream waterways?
2) *Available Space*: Does the park have available space for water flow and storage?
3) *Absorption*: Does the composition of the existing soils, water table, and underlying rock facilitate the absorption of significant amounts of water in the necessary amount of time?

Green Infrastructure Best Practices for Parks

A typical park already contains open-space areas that absorb stormwater and offer water quality, habitat, and esthetic benefits. Park infrastructure, such as parking lots, roads, buildings, playing fields, courts, and other artificial surfaces, can be built or retrofitted to treat stormwater or drain it to enhanced pervious surfaces. Many green infrastructure best practices can be implemented to manage stormwater, including the following:

- *Infiltration Beds*: Infiltration beds retain and infiltrate stormwater in a subsurface layer to reduce the volume of runoff entering storm sewers and increase groundwater recharge. The benefits include reduced runoff, improved water quality, effective use of space through layering of stormwater services below other park uses, reduced need for stormwater conveyance structures such as pipes and inlets, and reduced stormwater volumes in areas of high erosion or downstream damage. Some of the things that need to be taken into account when installing infiltration beds are that the design must be considered early in the design phase to account for the needs of the system in conjunction with surface land use, an uncompacted subsurface with well-drained soils is required for infiltration, and the beds' use may be limited by regulations in areas of soil contamination.
- *Rain Gardens and Bioretention*: Rain gardens are shallow planted areas that retain small amounts of stormwater for 12–48 hours. They manage small storm events through infiltration and evapotranspiration via the plants they contain. They can also be designed for temporary storage, known as bioretention. Rain gardens are highly versatile and can be shaped and adapted to fit a number of constrained surroundings, including playgrounds and court sports, lawn areas, traffic islands, and roadway swales. In large parks, rain gardens are typically designed as part of a larger stormwater management system. They provide a variety of benefits in addition to reducing runoff and combined sewer overflows, including enhancing the resilience of the landscape, removing pollutants via settling, vegetation, and soils, providing a visually appealing landscape, and enhancing ecosystem services such as air quality improvement, habitat, and evaporative cooling.

- *Stormwater Planter Boxes*: Planter boxes are concrete or brick structures filled with soils and plants that can temporarily store runoff and provide treatment. They usually capture runoff from roofs and are suitable for urban areas with limited space. They may be underlined, for slow discharge to a storm sewer, or open-bottomed, to serve as infiltration systems, depending on the location and site conditions.
- *Porous Pavement*: Porous pavement consists of a porous (permeable) surface of asphalt, concrete, or pavers overlaying an open-graded stone storage/infiltration bed. Stormwater drains through the surface, is temporarily held in the voids of the stone bed, and then slowly drains into the underlying, uncompacted soils. The system is designed with an overflow outlet so that water does not rise to the pavement level. In addition to reducing runoff and combined sewer overflows, porous pavement has increased longevity compared to standard pavement because freeze–thaw cycles do not adversely affect its structural integrity. It can also reduce contaminants (e.g., total suspended solids) if built using the right base materials, as it filters the water before it enters the aggregate reservoir.
- *Green and Blue Roofs*: Green roofs can be established on both existing and new roofs, depending on the structural capacity of the building. Extensive green roofs have thin layers of soil with lightweight plants, while intensive green roofs support larger plants, including trees and shrubs. Blue roofs retain stormwater runoff on a rooftop, delaying it from entering the sewer system. They are a lower-cost alternative to green roofs.
- *Managing Rooftop Runoff*: Cisterns, rain barrels, and dry wells can be used to capture and store precipitation draining from buildings. These structures hold water for extended periods, enabling its reuse for irrigation, fire protection, or building services. Specifically:
  - Cisterns capture runoff from the rooftops of large buildings or other large impervious areas. They can be placed above- or belowground and are readily available in many sizes to accommodate capacity needs.
  - Rain barrels are used to capture limited amounts of stormwater for later reuse (e.g., plant watering).
  - Dry wells are small, subsurface irrigation pits or chambers that are used to collect roof runoff from the downspouts of individual buildings. They reduce the volume of runoff entering the stormwater infrastructure system by disconnecting impervious roof systems.
- *Constructed Wetlands*: Constructed wetlands mimic the functions of natural wetlands to capture runoff, improve water quality, and provide habitats for wildlife. They are usually constructed by excavating, backfilling, grading, dyking, and installing water control structures to establish desired hydraulic flow patterns. If they are to be built on highly permeable soils and impervious compacted clay, a liner is usually installed, and the original soil placed over the top. Wetland vegetation is then planted or allowed to establish naturally (NYC Parks 2010; NRPA 2017; US EPA 2017; Brears 2018).

Multiple Benefits of Green Infrastructure in Parks

There are multiple benefits of using green infrastructure in parks to manage stormwater runoff, including:

- *Preventing Localized Flooding*: In addition to damaging infrastructure and property and threatening human life and well-being, frequent flooding can cause anger among local communities and lead to the exodus of businesses. In some situations, the most cost-effective solution may be to remove structures and turn the area into parkland that periodically inundates.

- *Reducing Stream Pollution by Holding and Slowing Water*: Bioswales, rain gardens, and vegetated buffer strips can detain stormwater so that precipitation does not wash out large quantities of sediment and pollutants. The swales also filter out chemicals, animal waste, oils, and heavy metals, improving the quality of runoff.
- *Recharging Groundwater*: Allowing rain to percolate into the soil is essential, as groundwater is a significant source for drinking and irrigation.
- *Providing Sewer Ratepayers with More Visible and Tangible Benefits*: Though gray infrastructure can successfully transport and treat stormwater, it provides only the unitary benefit of pollution conveyance. In contrast, green infrastructure in parks provide multiple benefits that are visible to the public.
- *Enhancing Recreational Value*: Green infrastructure can be used to create or enhance park amenities (e.g., walking trails). Restoring degraded areas can provide wildlife habitat and viewing points, as well as opportunities for outdoor education.
- *Reducing Maintenance Costs*: Green infrastructure can be used to help reduce park maintenance. Stormwater utility funds can improve drainage, reduce erosion, and eliminate standing water. The health of vegetation and the maintenance burden can be improved through better drainage.
- *Providing Educational Opportunities*: Placing signs where green infrastructure is used raises public awareness of the importance of measures to manage stormwater volume and reduce pollution (TPL 2016; Raja Segaran et al. 2014; US EPA 2017; Feldman et al. 2019).

Furthermore, park agencies can improve ecological function and reduce costs by collaborating with stormwater utilities. In particular:

- *Contributing to Improved Park Hydrology*: When water rushes into parks, it can not only flood playing fields and undermine trees but also wash out streams and transport sediment downstream. Since every green infrastructure solution (e.g., a new swale) reduces the burden on waterways and wastewater treatment plants, it may be cost-effective for stormwater utilities to help pay for these improvements on parkland. Furthermore, these features are attractive, ecological alternatives to the placing of drains in recreational fields.
- *Lowering Irrigation Costs*: Many park agencies must pay for water; even if the water bill is paid for by a government agency, it is still a cost to taxpayers. Some park agencies must also pay a stormwater or drainage fee to a sewer agency based on the size of their property holdings and the percentage of their impermeable land. Therefore, the less water is sent down the drain, the lower the water and stormwater fees (TPL 2016).

Creating New Parks and Renovating Existing Parks with Green Infrastructure

New parks can be specially designed to manage stormwater using green infrastructure. For instance, the land can be regraded to an optimal slope and shape, a sub-base installed, engineered soil added to increase absorption, and plant materials explicitly selected to manage stormwater. Meanwhile, modifying a current park to manage stormwater using green infrastructure is financially beneficial as the land is available at no cost. However, modifying a park's existing features and uses may result in resistance from current users who oppose new design or management changes (TPL 2016).

There are many water-related considerations when creating a new park or renovating an existing one to manage stormwater, including surface and stormwater management, potable water needs, irrigation, and wastewater disposal. The presence or absence of water and related elements will affect decisions on landscape design, the ability to infiltrate stormwater, and the ability to implement green infrastructure best practices. As such, design decisions for all water needs and water-related issues must be informed by data gathered as part of the site inventory and analysis (Table 8.1) (NYC Parks 2010).

Table 8.1 Site inventory and analysis to inform park design decisions.

Site Inventory Aspect	Implications
History: A review of historic site conditions will identify previous drainage patterns and soil and vegetation conditions	
Topography: What was the original topography of the site? Was the area originally upland or lowland? Was it once a wetland or floodplain?	Areas of the filled floodplain or high-water table are likely to experience wet conditions continually If the area was excavated, topsoil might be lacking, with bedrock affecting site design and best practice selection If the site was once a stream, the original stream is likely to be buried in a pipe, with groundwater flowing along the pipe bedding
Water Flow: What water has been present at or flowed through the site?	Streams are often buried, but the geological conditions that led to their formation and movement are likely still to exist Groundwater levels may be high or moving through the site along the former stream channel Designers may wish to incorporate components that remind users of "buried streams" Visible disturbances of hydrological patterns may indicate contamination
Utilities: Are there utilities related to past uses?	Old utilities (e.g., water or sewer lines) that are no longer in use may affect the construction and performance of green infrastructure
Context: Each site exists in relation to myriad systems that influence its performance	
Drainage Area: What is the watershed or drainage area of the site? Where is the site located within its larger drainage area?	The site's location in the watershed must be identified Sites located in upper watersheds are less likely to have stormwater problems or flooding from off-site and are more likely to have better soils for infiltration Understanding watershed or drainage area issues (e.g., eroding stream banks, nutrient enrichment, and habitat loss) can inform green infrastructure design selection and contribute to watershed restoration

Table 8.1 (Continued)

Site Inventory Aspect	Implications
Neighboring Sites: What are the neighboring sites and conditions?	Depending on adjacent land uses, the design of stormwater best practices may involve extensive structures or primarily soil and vegetation strategies. The site may be able to connect water elements (e.g., stream buffers) or provide a missing water component The history of neighboring sites must be reviewed, to identify sources of potential contamination of the project site
Neighboring Structures: Are there adjacent building foundations, subway tunnels, or utility vaults?	The design of green infrastructure must take into consideration potential impacts on adjacent structures
Sewers: Is the site served by combined sanitary and storm sewers or by separate storm sewers?	Calculating and planning to capture runoff volume is essential for water quality in both combined and separated sewer systems. If volume reduction is not possible, delaying the flow of stormwater to combined sewers will help reduce the frequency and volume of combined sewer overflows into waterways
Tidal Conditions: Is the site or are the sewers impacted by tidal conditions? Where is the water table?	In older sites by waterfronts, streams and wetland flows will often have been placed in pipes, which may have included tidal gates. If such gates were not installed or are malfunctioning, water levels will vary, which will determine the type and placement of green infrastructure selected for the site
Zoning and Codes: How do the zoning and regulatory constraints relate to water?	The site may be subjected to floodplain high-water levels, a regulatory floodplain, or coastal zone setbacks Local codes may require downspouts and storm sewers to be connected directly to a public sewer. However, it may be possible to design green infrastructure that manages water on-site and only overflow to the public system
Conditions: Numerous conditions above- and belowground will help establish physical, legal, and esthetic parameters for developing a park project	
Suggested Conditions for Mapping: The following information, when applicable, should be added to the base plan by the designer	Hydrologic features and flow paths (e.g., existing streams and wetlands, existing springs, and areas of flow drainage) Mapped regulatory floodplains and areas of observed flooding Buried streams and seeps
Subsurface Conditions: The following information, when applicable, should be added to the base plan by the designer	Areas of well-draining soils or wet soils Any relevant information from soil tests, subsurface soil, water, and geologic information All existing utilities
Other: The following information, when applicable, should be added to the base plan by the designer	Supplemental information from the review of historical or other documents, areas of concern

(*Continued*)

Table 8.1 (Continued)

Site Inventory Aspect	Implications
Site Analysis: During site analysis, qualitative judgments are made about the capacity of the site	
Absorptive Capacity: What is the absorptive capacity of the landscape?	Bulk density testing should be conducted to determine soil compaction
	Percolation tests should be conducted
	Soil analysis should be conducted to determine natural infiltration rates. This can also determine if and what soil modifications are necessary to create an engineered soil mix that can absorb and filter water and support the plants integral to system function
	Surrounding infrastructure, including buildings, tunnels, and utilities, should be assessed to ensure it is not compromised by infiltration areas
Surface Permeability: What impervious surfaces exist? Where is it feasible to reduce or replace them with pervious ones?	Street and footpath width requirements should be examined and reduced where possible, keeping site usage in mind
	Parking requirements should be assessed, and it should be determined whether peak loads can be accommodated with overflow parking using pervious surfaces
	Replacement of asphalt or concrete pavements with porous pavements should be considered
	Paving demands should be examined to determine the minimum paving required to meet the need
Drainage: Can drainage be redirected to natural systems or green infrastructure?	It should be determined whether site drainage can be directed to adjacent blue belts, natural drainage systems, or other stormwater best practice locations
	Areas where there is erosion or overburdened infrastructure should be identified, and opportunities should be found to detain or redirect the water

Source: Modified from NYC Parks, "High Performance Landscape Guidelines: 21st Century Parks for Nyc".

Case 8.1 Kallang River @ Bishan-Ang Mo Kio Park

Kallang River @ Bishan-Ang Mo Kio Park in Singapore was a collaboration between Singapore's Public Utilities Board (PUB) and the National Parks Board aimed at turning a 2.7 kilometer concrete canal into a picturesque river rich in biodiversity. Under the Active, Beautiful, Clean Waters Programme, a combination of plants, natural materials, and civil engineering techniques were introduced to soften the edges of the waterway, giving it a natural appearance and preventing soil erosion. The river channel was designed on a floodplain concept, linking it to a network of drains in the city. During dry weather, the flow of water is confined to a narrow stream in the middle of the river. During heavy storm events, the adjacent park area doubles up as a conveyance channel, carrying the rainwater gradually downstream. Students and teachers from Raffles

Institution created an interactive learning trail highlighting the features and rich biodiversity of the site. On this experiential trail, participants learn about Singapore's water management, as well as the transformation of the canal into a meandering river. ITE College Central and Raffles Institution also collaborated to develop a preschool learning trail for children, so that they can learn to appreciate Singapore's water sources (PUB n.d.).

Source: PUB, "Kallang River @ Bishan-Ang Mo Kio Park," https://www.pub.gov.sg/abcwaters/explore/bishanangmokiopark.

Urban Forests

Urban forests can be defined as networks or systems comprising all woodlands, groups of trees, and individual trees located in urban and peri-urban areas. They include forests, street trees, trees in parks and gardens, and trees in derelict locations. Urban forests bridge the rural and urban areas. Urban and peri-urban forestry (UPF) is the practice of managing urban forests to ensure their optimal contribution to the physiological, sociological, and economic well-being of urban societies. It is an integrated, interdisciplinary, participatory, and strategic approach to planning and managing forests and trees in and around cities. It involves the assessment, planning, planting, maintenance, preservation, and monitoring of urban forests, and it operates at scales from single trees to landscapes.

Urban Forest Types

The main urban forest types are:

- *Peri-Urban Forests and Woodlands*: Forests and woodlands surrounding towns and cities that can provide goods and services such as wood, fiber, fruit, other non-wood forest products, clean water, recreation, and tourism.
- *City Parks and Urban Forests (>0.5 ha)*: Large urban or district parks with a variety of land cover and at least partly equipped with facilities for leisure and recreation.
- *Pocket Parks and Gardens with Trees (<0.5 ha)*: Small district parks equipped with facilities for recreation/leisure, and private gardens and green spaces.
- *Trees on Streets or in Public Squares*: Linear tree populations, small groups of trees, and individual trees in squares and parking lots on streets.
- *Other Green Spaces with Trees*: Urban agricultural plots, sports grounds, vacant lands, lawns, riverbanks, open fields, botanical gardens, and so forth.

Urban Forests and Water

Sustainable watershed management is a crucial aspect of sustainable urban development. While cities only occupy 2% of the earth's terrestrial surface, they account for 75% of all residential and industrial water use. Healthy urban and peri-urban watersheds provide good-quality water for residential, industrial, and agricultural uses, as well as a range of ecosystem services. Well-managed and healthy urban forests can contribute to the sustainable management of water and water resources in urban and peri-urban watersheds.

They can help clean, save, and store water and reduce the risk of water-related disasters. Specifically, in addition to increasing the availability of water, urban forests can increase its quality by intercepting air pollutants, reducing sediment, and filtering rainwater. They can help minimize damage from runoff in urban and peri-urban environments. Furthermore, they can increase soil filtration, reducing the severity of flooding events (FAO 2016; Livesley et al. 2016; Berland et al. 2017; Kuehler et al. 2017; Van Stan et al. 2018; Escobedo et al. 2019; Phillips et al. 2019). A set of guidelines on how urban forests can be effective in managing water and water resources are provided in Table 8.2 (FAO 2016).

Table 8.2 Guidelines for maximizing the contributions of urban forests to sustainable management of water and water resources in watersheds.

Action	Description
Policy and legal framework	Ensure the role of UPF is adequately addressed in policies and laws directed at the minimization and remediation of water contamination and pollution
	Adopt policies to protect and conserve watersheds using green infrastructure approaches such as forest conservation and the retention of riparian vegetation along waterways
	Protect forests in watersheds from urban development and damage through adequate land-use planning
	Establish policies and regulations recognizing the decisive role that urban forests can play in the treatment and disposal of industrial wastewater
Planning, design, and management (protecting watersheds for water supply and water quality)	Use proactive planning approaches to reduce soil erosion and control sediment flows
	Conduct field assessments of existing forests in watersheds and identify potential restoration opportunities
	Adopt silviculture approaches aimed at maintaining and improving water quality
	Restore degraded watersheds through tree planting, agroforestry, and natural regeneration to improve watershed functioning
	Establish riparian forest buffer zones to protect streams, lakes, and other wetlands from disturbances and encroachment
	Use innovative tree-based approaches to help reduce water consumption, improve water quality, and recycle wastewater in urban and peri-urban environments
Increasing resilience to flooding events/ stormwater runoff	Assess and monitor flooding and stormwater runoff in urban and peri-urban areas
	Increase the percentage of permeable surfaces and tree cover, especially in urban areas particularly affected by flooding and stormwater runoff events
	Apply, as appropriate, green infrastructure approaches (e.g., forested bioswales, permeable pavements, green roofs, green streets, wooded wetlands, rain gardens, bioretention, bioinfiltration, forested filter strips, and linear stormwater trees) to mitigate the impacts of stormwater runoff

Source: Modified from FAO, "Guidelines on Urban and Peri-Urban Forestry".

Multiple Benefits of Urban Forests

In addition to regulating the urban hydrological cycle, urban forests provide a wide range of benefits, as listed in Table 8.3 (Sanesi et al. 2011; FAO 2016; Grote et al. 2016; Livesley et al. 2016; Endreny et al. 2017).

Management of Trees and Ecosystem Services

Urban forests provide ecosystem services that contribute to human health, livability, and sustainability. For example, in addition to reducing stormwater runoff, they also sequester carbon, remove air pollutants, and create green space that encourages

Table 8.3 Multiple benefits of urban forests.

Urban Issue	Benefits
Flooding	Mitigate stormwater runoff and reduce flooding
Stormwater costs	Reduce the volume of stormwater treated at wastewater facilities
Limited water resources	Enable infiltration and the reuse of stormwater
Food security	Provide food, clean water, and wood fuel
Urban poverty	Create jobs and increase income
Soil and landscape degradation	Improve soil conditions and prevent erosion
Reduced biodiversity	Preserve and increase biodiversity
Air and noise pollution	Remove air pollution and buffer noise
Greenhouse gas emissions	Sequester carbon and mitigate climate change, improve the local climate, and build resilience
Extreme weather events	Mitigate the local climate and build resilience
Energy shortage	Save energy through shading/cooling and grow wood fuel
Heat island effect	Cool the built environment through shade and evapotranspiration
Limited accessible green space	Provide more accessible natural and green space
Public health	Improve the physical and mental health of residents
Limited recreational opportunities	Provide opportunities for recreation and environmental education
Exposure	Provide shelter
Lack of community and social cohesion	Provide distinctive places for formal and informal interaction

Sources: FAO, "Guidelines on Urban and Peri-Urban Forestry"; Livesley, McPherson, and Calfapietra, "The Urban Forest and Ecosystem Services: Impacts on Urban Water, Heat, and Pollution Cycles at the Tree, Street, and City Scale."; Giovanni Sanesi, Christos Gallis, and Hans Dieter Kasperidus, "Urban Forests and Their Ecosystem Services in Relation to Human Health," in Forests, Trees and Human Health, ed. Kjell Nilsson, et al.(Dordrecht: Springer Netherlands, 2011); T. Endreny et al., "Implementing and Managing Urban Forests: A Much Needed Conservation Strategy to Increase Ecosystem Services and Urban Wellbeing," Ecological Modelling 360(2017); Rüdiger Grote et al., "Functional Traits of Urban Trees: Air Pollution Mitigation Potential," Frontiers in Ecology and the Environment 14, no. 10 (2016).

recreation, socializing, and relaxation. The management of trees influences the delivery of these services as follows:

- *Size*: Ecosystem service delivery increases with tree size. Avoided stormwater runoff, carbon storage, and air pollution removal increase with trunk diameter. Rainfall and air pollution interception increases with greater canopy size and total leaf area.
- *Species*: Ecosystem service delivery varies among species. Characteristics linked to differences in ecosystem services delivery include the following:
  - Deciduous trees intercept less rainfall and airborne pollutants when leafless.
  - Trees with rough or flaky bark or rough or hairy leaf structures trap and retain more air pollutants than those with smooth bark and leaves.
  - Trees with multilayered branching and denser canopies intercept more rainfall than those with open canopies.
- *Condition*: Trees in poorer condition provide lower ecosystem services delivery. Poor condition impedes growth, reducing the capacity to intercept precipitation and airborne pollutants (Forest Research 2019).

Planning and Designing Urban Forests

In urban centers, healthy trees and forests require careful planning, design, and management to achieve their full economic, social, and ecological potential.

Planning Urban Forests

A critical viewpoint in the planning of urban forests is that the environment hosts the city. Planning involves both the development of open spaces and the revitalization of existing areas. The planning stage involves goal setting, data collection and analysis, forecasting, design, strategic thinking, and public consultation. Urban forests should be a priority in municipal planning strategies, with the objective of maximizing the multiple benefits provided by trees and green infrastructure. There are five critical steps in integrating urban forests in municipal planning processes:

1) *Addressing UPF in Urban Areas*: Urban plans should reflect local policies and provide a framework for implementing land-use regulations. They should address all aspects of green infrastructure. They should ensure that green spaces receive equal attention in the urban planning process as elements of the built environment and are viewed as crucial components of infrastructure, providing the city with ecosystem services and multiple co-benefits.
2) *Fostering Dialogue Between UPF and Other Planning Components*: Where possible, urban plans should establish meaningful links between urban forests and other planning components. For example, urban parks should be considered complementary to economic development plans, with parks hosting festivals and cultural events that draw tourists and strengthen local economies. Cross-referencing land-use elements can help identify risks, such as those posed to urban forests by planned developments, enabling actions to be devised to minimize them. Opportunities should be created to bring the staff of relevant departments and stakeholders together in the planning process to draft strategic, sectoral, and operational plans collaboratively.

3) *Including a UPF Evaluation Checklist or Guidelines Among the Technical and Legislative Norms of City Development Strategies*: A checklist of indicators, actions, and achievements can make it easier to discuss, evaluate, and incorporate the services of urban forests in city development strategies. The most successful systems for monitoring urban forest plans are those that can be incorporated into standard maintenance activities.
4) *Taking an Adaptive Management Approach to Urban Forest Resources*: An adaptive management approach, including regular monitoring and evaluation, will ensure than urban plans respond rapidly and appropriately to the evolving needs of urban communities.
5) *Planning for the Long-Term Maintenance of Urban Forests*: Urban plans generally span several years, but trees, and forests, may live for centuries. Integrating urban forests in city planning requires the adoption of a long-term perspective on their management, maintenance, and conservation.

Designing Urban Forests

Designing urban forests involves creating sustainable living spaces that are beneficial to both humans and nature. Urban forest design can be applied to the creation of new forests and the redesign of existing ones, with the latter involving interventions that aim to improve existing green spaces in order to provide new services and facilities and enhance their sustainability. The redesign of existing forests can even be part of the redesign of a broader urban landscape, integrating them with architecture, transport, and planning to create urban ecosystems that contribute to climate change mitigation and adaptation.

All urban forest design processes should first identify suitable spaces. There are three main types of location for forests and trees in urban and peri-urban settings:

1) Trees in streets, squares, parking areas, and other "gray" spaces with sealed surfaces.
2) Trees in parks and other green spaces (e.g., continuous soil strips, properties, gardens, and commercial areas).
3) Stands, patches, and other groups of trees, which may be referred to as "woodlands," "woods," or "forests."

All potential stakeholders in a community should be consulted in urban forest design so that it adequately reflects their needs, requirements, and demands. The design process should be flexible to accommodate the outcomes of the consultation process. It is also essential that it promotes the social comfort of users by meeting the needs of the community, is compatible with the specific characteristics of the site, creates places where trees can thrive, delivers a full range of ecosystem services, and helps meet sustainable development goals.

A key aspect of urban forest design is ensuring that the right tree is in the right place. Effective design is based on a clear understanding of the environmental, infrastructural, and sociocultural characteristics of the site. Environmental characteristics include climate, soil, biology, and ecology. Infrastructure characteristics include the relationships with natural and artificial infrastructure. Sociocultural characteristics include community preferences. A well-designed urban forest will provide multiple benefits, including cooling homes in summer, reducing wind speeds, and improving the health of the natural environment. Effective design will also ensure the forest does not negatively affect infrastructure, such as buildings and power lines.

The design of urban forests has six dimensions. The social, functional, ecological, and economic dimensions interact with the perceptual dimension over the temporal dimension to provide ecosystem services and other benefits (Table 8.4) (FAO 2016).

Table 8.4 Dimensions of urban forest design.

Dimension	Description	Aspects
1. Perceptual	How people perceive the environment and experience places affects their relationship to urban forests	A good urban forest design should: • Emphasize the identify, structure, and meaning of the location • Create thematic spaces related to the specific needs of the community • Create or enhance a sense of place (the emotional bonds that people form with a place over time)
2. Sociocultural	Cultural, social, physical, and ecological elements and processes determine the preferences, perceptions, and uses of urban forests in a community	*Meeting the Needs of Users*: Landscapes should be designed to allow people to become familiar with green spaces at their own pace *Location*: The location of an urban forest significantly determines who enjoys it, how often, and at what times *Size*: The size of an urban forest influences the kinds of experiences a visitor has: 1) Small forests are refuges for contemplation and for children to engage in play and learning 2) Larger forests can accommodate multiple uses and may offer a "wilderness experience" 3) Vegetation structure provides a sense of enclosure or protection, but an open structure is usually preferred in urban settings 4) Multiple features, such as the presence of water, broad views, lawns, and spaces for relaxation, social gatherings, and sporting activities, all help ensure the success of an urban forest
3. Ecological	The design of urban forests should adopt ecological principles at various scales (macro-scale: municipality; meso-scale: neighborhood; micro-scale: individual building). This encompasses structural, species, and spatial diversity, as well as habitat connectivity	Urban forests can connect different habitats for migratory species, while their age, size, and composition have essential impacts on plant and animal communities. There are three ways of designing the ecological dimension of urban forests: 1) *Ecological Succession Approach*: Natural ecosystems develop unassisted 2) *Close-to-Nature Design*: There is a certain level of intervention and management as trees and forests develop, mimicking ecological processes 3) *Artificial Construction*: Urban forest elements are designed with little relation to natural ecosystems
4. Functional	Urban forests can perform many functions	In city environments, urban forests produce a wide range of economic, social, and environmental benefits. Balancing competing demands and taking a multifunctional approach is critical in the design process

Table 8.4 (Continued)

Dimension	Description	Aspects
5. Economic	The economic efficiency of urban forest design is determined by comparing the benefits derived from the forest with the costs it incurs. For example, the creation of an urban forest may provide certain ecosystem services at a lower cost than conventional gray infrastructure	The economic dimension of UPF design has four elements: 1) The economic value of the ecosystems it generates 2) The urban forest's perceived value (e.g., the value of its acting as a noise barrier) 3) The role of an urban forest in providing income, employment, and subsistence 4) The means of financing the design and management of an urban forest, including the opportunity cost of devoting an area of land to public green space
6. Temporal	Urban foresters should aim to design forests that can meet current needs and address current pressures, yet be resilient and adaptable to change over time	A vital element of the resilience and adaptability of urban forests is structural diversity in terms of age, spatial profile, and species distribution. Structural diversity can be achieved by quickly planting appropriate fast-growing pioneer trees and shrubs in addition to slower-growing trees

Design and Management for Goods and Ecosystem Services

Overall, urban forest designs usually envisage the provision of a wide range of goods and ecosystem services. Urban forest designers should find synergies with managers of other natural resources and other land uses in urban and peri-urban areas. Effective designs can help create smart cities by promoting the integration of green infrastructure, townscapes, sports areas, forestry, agroforestry, agriculture, and so on. Designers should be aware of the increasing focus on green infrastructure solutions to significant challenges in urban environments.

Sources: Modified from FAO, "Guidelines on Urban and Peri-Urban Forestry".

Case 8.2 City of Vancouver's Urban Forest Strategy

The City of Vancouver, Canada is aiming to plant 150 000 trees by 2020 as part of its Urban Forest Strategy. The purpose of this strategy is to protect, plant, and manage trees in order to create a diverse, resilient, and beautiful urban forest on public and private lands across the city. Vancouver's urban forest includes every tree in the city (on streets, parks, public spaces, and properties). The strategy has 12 guiding principles, as listed in Table 8.5, while its specific goals and targets are as follows.

Urban forest goals:

- Protect the urban forest during development.
- Plant trees to grow the urban forest.
- Manage trees for health and safety.
- Engage citizens in the urban forest.
- Monitor the status and condition of the urban forest.

Table 8.5 Guiding principles of the City of Vancouver's urban forest strategy.

Principle	Description
1) Protect	Protect the city's existing urban forest
2) Beauty	Create beautiful urban landscapes
3) Biodiversity	Enhance habitat and support biodiversity
4) Planning	Select the right tree for the right place
5) Equity	Distribute ecosystem services equitably
6) Climate change	Mitigate and adapt to climate change
7) Health and safety	Support healthy and safe trees
8) Livable city	Contribute to a healthy city
9) Culture and identity	Celebrate Vancouver's cultural and natural identity
10) Stakeholder engagement	Collaborate with a broad range of partners
11) Science-based	Use sound science
12) Measure	Measure progress

Source: Modified from City of Vancouver, "Vancouver's Urban Forest Strategy," https://vancouver.ca/home-property-development/urban-forest-strategy.aspx.

Urban forest targets:

- Plant 150 000 trees between 2010 and 2020.
- Increase the urban forestry canopy to 22% by 2050.
- Restore or enhance 25 hectares of natural areas, including forests, by 2020.
- Double street tree density in below-average blocks of the Downtown Eastside, Marpole, and other priority neighborhoods.

The City of Vancouver is developing a green infrastructure strategy to improve and protect its water quality and proactively prepare for climate change. Trees manage water naturally, and so the urban forest will play a significant role in the city's Rain City Strategy, which has the target of capturing and cleaning 90% of rainfall on both public and private land, while at the same time:

- Improving and protecting the city's water quality.
- Increasing the city's resilience through sustainable water management.
- Enhancing the city's livability by improving natural and urban ecosystems.

At the end of 2017, over 102 000 trees were planted, 55% on public lands (streets and parks) and 45% on private ones, including backyards and development sites.

Tree planting on public lands is divided into three components:

1) *Street Trees*: Around 1500 street trees have been planted annually on publicly managed rights of ways and boulevards. Many of the readily available planting spots have been filled, and new planting is targeting hard-to-plant sites with poor soils and replacement planting of unhealthy street trees.

2) *Ornamental Park Trees*: Between 300 and 500 ornamental park trees are planted in Vancouver parks each year, in order to improve park esthetics, provide shade around playgrounds, and replace dead or unhealthy trees.
3) *Forest Restoration*: Over 70% of new trees planted on public lands (around 7000 trees per year) are native tree species, planted to support the city's ecological network and increase biodiversity.

Private-land planting has averaged around 9000 trees per year since 2015. Its main components are:

1) *Spring and Fall Tree Sales and Nursery Sales*: The Park Board provides incentives for homeowners to plant trees on residential properties. The trees are typically 5 gallons in size and sold for $10. These sales account for more than half (about 5000) of the trees planted on private lands annually. A new program allows Vancouver residents to purchase trees at local nurseries and receive a rebate of $20 off the retail purchase price.
2) *Development Trees*: The city's planning staff review development applications and work with landowners, consultants, and developers to protect existing trees and plant replacements, and to increase tree cover if possible. Around 3800 trees are planted as part of development applications annually.
3) *Additional Private Land Trees*: It is estimated that about 500 trees per year are planted on private property but not captured as part of the Park Board tracking process. These trees are purchased from nurseries throughout the region, at plant sales, or from other sources (City of Vancouver n.d.).

Source: City of Vancouver, "Vancouver's Urban Forest Strategy," https://vancouver.ca/home-property-development/urban-forest-strategy.aspx. © City of Vancouver

Peri-Urban Forests (Green Belts)

Urban forests can extend to green belts around cities that buffer waterways and regulate development. They can even extend to the acquisition and management of lands to preserve urban watersheds and protect drinking water supply and quality (Brears 2020).

Case 8.3 Vienna's Green Belt

Vienna's green belt, which is composed of woodland and meadows plus a mountain road, was established in 1905, making it one of the oldest urban green belts in the world. Initially amounting to 6000 hectares, it expanded over 100 years to encompass 30% of the entire urban area, with parts of the city designated as green belt protection areas and green belt agricultural protection areas. The green belt protects Vienna's water supply, with residents supplied with 400 000 cubic meters of spring water per day from the areas around Schneeberg, Rax, Schneealpe, and Hochschwab. On the centennial of the green belt's founding, the Municipal Council adopted the Vienna Green Belt plan to encompass not only all existing and planned green belt protection areas, but also major parks and the Central Cemetery – in total, 19 260 hectares (City of Vienna 2007; Vienna Tourism Board n.d.).

Sources: Vienna Tourism Board, "The Vienna Woods," https://www.wien.info/en/sightseeing/excursions/vienna-woods; City of Vienna, "Vienna Environmental Reports," https://www.wien.gv.at/english/environment/protection/reports/

References

Berland, A., S.A. Shiflett, W.D. Shuster, A.S. Garmestani, H.C. Goddard, D.L. Herrmann, and M.E. Hopton. (2017). "The Role of Trees in Urban Stormwater Management." *Landscape and Urban Planning* 162: 167–177.

Brears, R.C. (2018). *Blue and Green Cities: The Role of Blue-Green Infrastructure in Managing Urban Water Resources.* London: Palgrave Macmillan.

Brears, R.C. (2020). *Nature-Based Solutions to 21st Century Challenges.* Oxford: Routledge.

City of Vancouver. (n.d.). "Vancouver's Urban Forest Strategy." https://vancouver.ca/home-property-development/urban-forest-strategy.aspx (accessed October 9, 2020).

City of Vienna. (2007). "Vienna Environmental Reports." https://www.wien.gv.at/english/environment/protection/reports/ (accessed October 9, 2020).

Endreny, T., R. Santagata, A. Perna, C. De Stefano, R.F. Rallo, and S. Ulgiati. (2017). "Implementing and Managing Urban Forests: A Much Needed Conservation Strategy to Increase Ecosystem Services and Urban Wellbeing." *Ecological Modelling* 360: 328–335.

Escobedo, F.J., V. Giannico, C.Y. Jim, G. Sanesi, and R. Lafortezza. (2019). "Urban Forests, Ecosystem Services, Green Infrastructure and Nature-Based Solutions: Nexus or Evolving Metaphors?" *Urban Forestry & Urban Greening* 37: 3–12.

FAO. (2016). "Guidelines on Urban and Peri-Urban Forestry." http://www.fao.org/3/a-i6210e.pdf (accessed October 9, 2020).

Feldman, A., R. Foti, and F. Montalto. (2019). "Green Infrastructure Implementation in Urban Parks for Stormwater Management." *Journal of Sustainable Water in the Built Environment* 5, no. 3: 05019003.

Forest Research. (2019). "Understanding the Role of Urban Tree Management on Ecosystem Services." https://www.forestresearch.gov.uk/research/understanding-role-urban-tree-management-ecosystem-services/ (accessed October 9, 2020).

Grote, R., R. Samson, R. Alonso, J.H. Amorim, P. Cariñanos, G. Churkina, et al. (2016). "Functional Traits of Urban Trees: Air Pollution Mitigation Potential." *Frontiers in Ecology and the Environment* 14, no. 10: 543–550.

Kuehler, E., J. Hathaway, and A. Tirpak. (2017). "Quantifying the Benefits of Urban Forest Systems as a Component of the Green Infrastructure Stormwater Treatment Network." *Ecohydrology* 10, no. 3: e1813.

Livesley, S.J., E.G. McPherson, and C. Calfapietra. (2016). "The Urban Forest and Ecosystem Services: Impacts on Urban Water, Heat, and Pollution Cycles at the Tree, Street, and City Scale." *Journal of Environmental Quality* 45, no. 1: 119–124.

NRPA. (2017). "Green Infrastructure in Parks." https://www.nrpa.org/contentassets/0e196db99af544bbba4f63f480c1316b/gupc-resource-guide.pdf (accessed October 9, 2020).

NYC Parks. (2010). "High Performance Landscape Guidelines: 21st Century Parks for NYC." https://www.nycgovparks.org/sub_about/sustainable_parks/design_guidelines.pdf (accessed October 9, 2020).

Phillips, T.H., M.E. Baker, K. Lautar, I. Yesilonis, and M.A. Pavao-Zuckerman. (2019). "The Capacity of Urban Forest Patches to Infiltrate Stormwater Is Influenced by Soil Physical Properties and Soil Moisture." *Journal of Environmental Management* 246: 11–18.

PUB. (n.d.). "Kallang River @ Bishan-Ang Mo Kio Park." https://www.pub.gov.sg/abcwaters/explore/bishanangmokiopark (accessed October 9, 2020).

Raja Segaran, R., M. Lewis, and B. Ostendorf. (2014). "Stormwater Quality Improvement Potential of an Urbanised Catchment Using Water Sensitive Retrofits into Public Parks." *Urban Forestry & Urban Greening* 13, no. 2: 315–324.

Sanesi, G., C. Gallis, and H.D. Kasperidus. (2011). "Urban Forests and their Ecosystem Services in Relation to Human Health." In *Forests, Trees and Human Health*, edited by K. Nilsson, M. Sangster, C. Gallis, T. Hartig, S. de Vries, K. Seeland, and J. Schipperijn. Dordrecht: Springer, pp. 23–40.

TPL. (2016). "City Parks, Clean Water: Making Great Places Using Green Infrastructure." https://www.tpl.org/sites/default/files/City%20Parks%20Clean%20Water%20report_0.pdf (accessed October 9, 2020).

US EPA. (2017). "Green Infrastructure in Parks: A Guide to Collaboration, Funding, and Community Engagement." https://www.epa.gov/sites/production/files/2017-05/documents/gi_parksplaybook_2017-05-01_508.pdf (accessed October 9, 2020).

Van Stan, J.T., S.J. Underwood, and J. Friesen. (2018). "Urban Forestry: An Underutilized Tool in Water Management." In *Advances in Chemical Pollution, Environmental Management and Protection*, edited by J. Friesen and L. Rodríguez-Sinobas. Philadelphia, PA: Elsevier, pp. 35–61.

Vienna Tourism Board. (n.d.). "The Vienna Woods." https://www.wien.info/en/sightseeing/excursions/vienna-woods (accessed October 9, 2020).

9

Water Bodies

Introduction

Local and regional authorities are generally responsible for land use decisions and have an important role in assessing environmental impacts and protecting, conserving, and enhancing natural ecosystems, including water bodies. Incorporating green infrastructure that decreases flood risk or improves water quality into related plans and strategies can overcome the fragmentation of habitats and preserve or restore ecological connectivity, enhance ecosystem resilience, and therefore ensure the continued delivery of ecosystem services while providing healthier environments and recreational spaces for people to enjoy (Miles and Band 2015; ESPON 2018; Golden and Hoghooghi 2018). This chapter will discuss various green infrastructure strategies that can be implemented for these purposes, including retention ponds and riparian buffers, as well as river engineering, floodplain restoration, wetland restoration, and green corridor initiatives.

Retention Ponds

A retention pond is a permanent pool of water with capacity to capture and slowly release additional water over a period of time. The permanent pool is replaced, in part, with stormwater during runoff events. Retention ponds are effective in removing suspended solids, organic matter, and metals through sedimentation and soluble pollutants such as dissolved metals and nutrients through biological processes. They require groundwater or a dry-weather base flow if the permanent pool elevation is to be maintained year-round. In addition to water quantity and water quality benefits, retention ponds create wildlife and aquatic habitat, as well as recreational, esthetic, and open space opportunities. They represent a cost-effective best management practice for larger tributary watersheds (Urban Drainage and Flood Control District 2010).

Regional Water Security, First Edition. Robert C. Brears.

Case 9.1 Constructed Stormwater Management Ponds in the City of London, Ontario

The City of London, Ontario has constructed stormwater management ponds to collect rainfall and surface water runoff and release it at a controlled rate. A single pond can provide erosion and flooding control and enhance water quality. In particular, ponds collect surface water runoff from rooftops, lands, and roads. The solids in the water settle to the bottom of the pond as sediment, while plants and bacteria help remove pollutants. Cleaner water then leaves the pond and makes its way to the Thames River. To ensure the ponds function as designed, the City of London has established the Adopt-A-Pond program, which encourages community groups, residents, and local businesses to get involved in the preservation of the city's stormwater management facilities. Participants volunteer their time to help maintain, beautify, enhance, and protect a pond for all to enjoy. The city meanwhile provides the materials and support necessary to complete the twice-annual clean-up activities. As part of the program, pond adopters agree to:

- Host a minimum of two clean-up days per year, spring and fall, for at least 2 years.
- Consider hosting additional events and clean-ups throughout the year.
- Observe and report any hazards.
- Remove graffiti.
- Paint yellow fish by storm drains as a visual reminder that only rainwater belongs there.
- Deliver information handouts to homeowners.
- Record and report volunteer hours and activities.

Meanwhile, the city agrees to:

- Provide the tools required for a successful clean-up day, including garbage bags.
- Help with any inquiries, assistance, or guidance participants may need.
- Erect an Adopt-A-Pond sign with the group's name on it, recognizing their commitment to the pond's maintenance.
- Assist with the purchase of trees, shrubs, and flowers and provide guidance and expertise on how to plant.
- Remove filled garbage bags from around the pond following clean-up days (City of London n.d.).

Source: Based on City of London, "Stormwater Management Ponds," https://www.london.ca/residents/Sewers-Flooding/stormwater/Pages/Living-Near-Stormwater-Management-Ponds.aspx.

Riparian Buffers

Riparian buffers are strips of vegetation established next to waterways in managed landscapes that are designed to capture runoff, nutrients, and sediment while restoring the natural aquatic environment. They typically range in width from 30 to 300 feet from the stream banks. The vegetation used can include grasses, shrubs, and a variety of trees. In addition to reducing stormwater runoff, riparian buffers reduce sediment and nutrient load, improve water quality, and provide wildlife habitat and recreational opportunities (University of Arkansas 2009).

Creating Riparian Buffers

Riparian buffers can be created passively (naturally) or actively. Passively creating a riparian buffer involves allowing the land to become naturally vegetated. Over time, plants will become established as seeds are introduced by birds, animals, water, and wind. Actively creating a riparian buffer, on the other hand, typically involves six steps:

1) *Site Assessment*: When evaluating the location for a riparian buffer, information such as land ownership, permits, site characteristics (climate, soils, and geology), utilities, and planned future land use should be gathered. The width of the buffer will depend on land availability, slope, soil infiltration capacity, soil quality, and site needs (e.g., what will the site be used for?).
2) *Planting Plan Development*: When developing a planting plan, the boundaries and constraints of the site need to be identified. Property boundaries, utilities, permit conditions, land use considerations, and budget, as well as unique land features such as springs, wet areas, and pedestrian and biking paths, should all be factored in. Stream banks should be planted with vegetation that can quickly colonize the environment to protect against erosion. Trees should be planted at a density of 800 per acre. The time of year needs to be ideal for each tree and grass species.
3) *Site Preparation*: Before the site can be planted, it must be prepared. Stream banks may need to be regraded before planting. Invasive species should be removed. Soil amendments may need to be added, such as lime and fertilizers. Proper noncompetitive ground cover can assist in establishing tree and shrub species while reducing maintenance and possible invasion from non-native species.
4) *Species Selection*: When selecting plants, factors such as water tolerance, project goals, natural succession, plant availability, and esthetics need to be considered. As riparian areas often experience wetting from floods, selected plants need to be tolerant of such conditions. Different vegetation types provide different benefits.
5) *Planting*: A wide variety of planting methods are available for different types of vegetation. For trees, bare-root seedlings are often used. Shrubs are commonly established from live stakes. Grasses can be sown by hand.
6) *Protection and Maintenance*: Once the riparian buffer is planted, it faces threats from livestock and wildlife browsing, insects, humans, and invasive species. The emerging vegetation should be monitored to ensure that undesired species are quickly removed. Fencing can prevent livestock and wildlife from grazing and accidental destruction of vegetation by mowing. Signage can be used to mark the boundaries of the riparian buffer (University of Kentucky 2010; Kiss et al. 2019).

Case 9.2 Auckland City Council's Waterway Protection Fund

Auckland City Council in New Zealand has created a Waterway Protection Fund for landowners, farmers, lifestyle block owners, and community and environmental groups in the Drury, Wairoa, and Waiuku catchment areas. Funding is available for projects that protect, enhance, or restore streams, waterways, wetlands, and riparian margins, that install fencing to exclude livestock, that help develop alternative water supplies, or that plant native plants along watercourses. It can provide up to 50% of project costs. Eligible projects are assessed for:

- Their projected environmental benefits, including in terms of water quality and aquatic ecological health.
- The type and significance of the environment and biodiversity to be protected.
- The length of riparian margin to be protected.
- The extent of community involvement.
- Connections with other landowners, with a view to forming partnerships and sharing knowledge.
- The feasibility and likelihood of successful completion (Auckland Council n.d.).

Source: Auckland Council, "Eligibility for the Waterway Protection Fund (Wpf)," https://www.aucklandcouncil.govt.nz/grants-community-support-housing/grants/regional-grants/Pages/eligibility-for-waterway-protection-fund.aspx.

Green Infrastructure and River Engineering

Green infrastructure refers to the network of natural and semi-natural features found among and between towns and cities. Rivers are part of this network, and green infrastructure promotes their conservation or restoration. This approach is fundamental to improving the morphology, ecosystem, and water quality of rivers, as well as contributing to enhancing their resilience to climate change.

There are two green infrastructure approaches available for river engineering: green measures, which use vegetative materials exclusively; and green-gray measures, which may involve the use of nonvegetative, nonbiodegradable components to provide structural stability and erosion resistance. In practice, individual green or green-gray measures are rarely used in isolation, but are typically just one component of a river engineering project. Meanwhile, associated river management techniques are actions (green and green-gray) that contribute to the creation of green infrastructure (Roca et al. 2017):

Green Measures

A variety of green measures are available to conserve and restore rivers, including the following:

- *Aquatic Vegetation*: Native plants well-adapted to living in a stream – that part of a channel that is perennially inundated – can be used to protect the toes of riverbanks by locally retarding near-boundary-flow velocities and providing root reinforcement to the bed and lower bank.
- *Coir Matting/Pallets*: These are biodegradable materials made from natural fibers that protect and stabilize the riverbank while allowing vegetation to grow through. Coir is the stiff fiber found on the outside of coconuts. It can be woven and pressed into many shapes, including matting of various thicknesses.

- *Faggots/Fascines/Brushwood*: Faggots and fascines are bundles of untreated brushwood bound together with biodegradable fibers and used to stabilize a riverbank's face or toe. The bundles slow the near-bank flow and trap fluvial sediment and organic matter.
- *Stakes*: Live stakes can be used to reinforce the bank face, promote vegetation regrowth, and anchor other measures in place until they grow and take root sufficiently to become self-supporting. Stakes are usually cut from stems and branches. The portion of the stake that is below ground will take root, and the exposed part will develop into either a tree or a bushy riparian plant. Dead stakes, also called posts, provide immediate effective bank reinforcement. In the long term, live stakes will provide superior performance, provided the resulting trees and shrubs are properly maintained.
- *Vegetation*: Grasses, herbaceous plants, and shrubs established on bare soil help to prevent surficial erosion by shielding the bank surface from raindrop impact (splash erosion), increasing infiltration and slowing runoff draining down the bank (sheet, rill, and gully erosion), retarding near-bank-flow velocities (fluvial erosion), and anchoring the subsoil through root reinforcement. Vegetation also takes up soil water, reducing the frequency of saturated conditions (most bank collapses are triggered by soil saturation) and generating negative pore water pressure that increases bank stability when the bank is unsaturated.
- *Willow Spiling*: Willow rods woven around stakes form a fence-like structure that is backfilled with soil to provide physical protection against erosion by fast-flowing water. The willow rods are live, but stakes may be alive or dead (i.e., posts). When installed, they can be used to deflect flow away from the bank temporarily and to promote near-bank deposition, which may change the course of the stream.
- *Woody Material*: Material sourced by felling trees can be used to deflect flows and promote sediment deposition at the toes of eroding banks. The wood can be installed in one of three forms: whole trees or tree trunks placed parallel to the flow at the bank toe; "root wads", which consist of tree trunks with the roots still attached that are pushed into the bank (trunk-first), leaving the roots exposed; and engineered log jams, which consist of large, expertly designed, interlocked wood structures that are kept in place by using some logs as vertically-driven piles. In these cases, the aim is to increase bank stability, shelter the bank surface, and increase bank roughness, all of which dissipates flow energy (Kiedrzyńska et al. 2015; Roca et al. 2017; Gunnell et al. 2019).

Green-Gray Measures

A variety of green-gray measures are available to conserve and restore rivers, including the following:

- *Geo Cell Systems*: These are three-dimensional earth-retaining structures that can be expanded and backfilled with material to stabilize the bank surface.
- *Vegetated Concrete Blocks*: These are articulated block systems that can adapt to irregular bank topography. Vegetation is applied in the form of live cuttings or grass plugs that are inserted through openings in the blocks into the soil beneath them.

- *Vegetated Gabions*: These are wire mesh baskets filled in situ with stones that can be used to stabilize the bank toe or piled up on top of each other to retain the bank at a steep angle. Woody vegetation is incorporated by inserting posts and poles through the basket during filling, which penetrate the underlying fill or intact soil. The spaces between the stones in the baskets can be filled with soil and seeded to promote vegetation growth.
- *Vegetated Reinforced Earth*: This includes grasses or live cut branches intermixed with soil that is wrapped in a natural fabric or geotextile. Vegetated reinforced earth can be used to protect the bank face.
- *Vegetated Reinforced Mattresses*: These are flexible mats that can be rolled out and fastened on to the bank slope. They can be made of natural or synthetic materials. They are designed to resist fluvial shear stresses and turbulent forces of traction and uplift. If all materials used are biodegradable, then this measure can be considered a green measure.
- *Vegetated Riprap*: This consists of layers of stones and boulders vegetated during construction using poles, brush-layering, or live-staking. It increases the stability of the bank or toe while establishing riparian growth.
- *Vegetated Rock Rolls*: These consist of a net material filled with cobble-sized stones and established vegetation. The rolls are usually used to stabilize the toe of the bank (Roca et al. 2017).

Associated River Management Techniques

Table 9.1 lists a variety of associated river management techniques (green and green-gray) that contribute to the creation of green infrastructure (Roca et al. 2017).

Table 9.1 River management techniques.

Technique	Description
Embankment removal	Involves the removal of flood embankments or the enlargement of set-back distances to provide more room for a river, particularly during flood events, by reconnecting it to its floodplain and side channels
Lowering/removal of weirs and culverts	Involves the modification or removal of weirs that are not functional or have fallen into disrepair and culverts that are undersized with regard to flood flows, sediment loads, and debris volumes as a result of climate change or land-use development in order to avoid risks of blockage and failure
Meander restoration	Consists of reconstructing meandering channels that have been straightened using green river engineering measures and techniques
Changing bed levels	Involves the modification of riverbed elevation by dredging or refilling (restoration of the bed elevation of an incised channel to reconnect a river to its floodplain is increasingly common globally)
Regrading of banks (or reprofiling)	Involves the excavation or backfilling of a stream bank to an appropriate slope
Widening/narrowing	Involves the modification of channel width by dredging, reprofiling of the banks, or application of in-stream structures to narrow the channel. This can increase the conveyance capacity of the channel by widening it, or it can locally increase flow velocities to promote habitat diversity by narrowing it

Case 9.3 Riverlands Work on Holnicote, Somerset, United Kingdom

The Riverlands work on Holnicote is the first time a river system has been restored to "Stage 0" in the United Kingdom, with its natural processes and habitats recovered. The initial project stretches across 10 acres of land and involves a tributary of the River Aller. Work is underway to revert the tributary to its original course before human interference. The pilot project will use earth-moving equipment to allow natural flow, sediment, and biological processes to develop a fully connected, stream-wetland system. After this work is completed, habitat restoration will be fast-tracked by using wood debris and key plant species to help develop more hydrological and ecological diversity on the site. This will benefit a host of plant and animal species, including 300 water voles released by the National Trust over the past 12 months. If the pilot project is successful, the same methods will be applied across 33 acres on the River Aller itself (National Trust n.d.).

Source: Based on National Trust, "River Restoration Work on Holnicote," https://www.nationaltrust.org.uk/holnicote-estate/features/river-restoration-work-on-holnicote

Floodplain Restoration

Floodwaters move much more slowly on floodplains, as they are usually flat and vegetated. As such, floodplains can act as reservoirs that temporarily store floodwater, slowly releasing it back to the river as the flood recedes. Floodplain restoration is the return of a floodplain's ecosystem to its natural conditions and functions before it underwent development. In addition to reducing flood levels and risk, this can improve water quality, provide a habitat for fish and wildlife, offer recreational opportunities, and provide erosion control (Blackwell et al. 2006; FEMA 2017).

Floodways or Flood Bypasses

Floodways or flood bypasses are portions of a river's historic floodplain that, during major flooding events, are reconnected to the river and become inundated. They act as relief valves in two ways:

- *Storage*: They can detain and store water, functioning similarly to a flood-control reservoir. In flood management, using a floodway for storage is known as "peak shaving" – reducing the height of the flood peak at some downstream point.
- *Conveyance*: They can provide conveyance, increasing the cross-sectional area available to move floodwaters safely through a stretch of river. In this way, they also reduce "backwater flooding," caused by the "piling up" of floodwaters at and behind a bottleneck. The vegetation within a bypass can influence its hydraulic roughness and ability to convey floodwaters, and so some bypasses are managed for vegetation with low roughness (Opperman 2014).

Other green infrastructure measures that can be implemented to reconnect floodplains include:

- *Levee Setback*: Moving levees back from the river channel can reconnect rivers to floodplains while still providing flood protection to areas behind them.

- *Levee Removal or Breaching*: Where levees are no longer needed to protect a floodplain from inundation, they can be breached in segments or removed entirely.
- *Floodplain Terrace Installation*: Where flows cannot reach the floodplain, a lower level can be excavated to offset the vertical difference.

Floodplain Restoration Activities

Dynamic river–floodplain processes generate habitat and diversity over time. A variety of restoration activities can be undertaken when land management practices interfere with the natural development of the floodplain over time or where floodplain areas are degraded to the point that intervention is required to stimulate natural processes:

- *Passive Restoration*: Landscaping, mowing, or clearing of vegetation in floodplains can interfere with natural ecosystem succession and reduce floodplain surface diversity. Cooperation with land managers helps ensure restored floodplain areas are managed to support natural ecosystem functions.
- *Secure Sediment Supply*: Flow-regulated rivers are usually starved of sediment, which becomes trapped behind dams in reservoirs. Continual sediment supply is necessary to establish new floodplain features, promote the recruitment of riparian vegetation, and prevent channel incision. Organized flushing of trapped sediment can help return it to the ecosystem.
- *Enhanced Woody Debris Recruitment*: Log jams and root wads create complex habitats on floodplains and side channels. Woody debris can also increase sediment storage in-stream and aggrade streambeds. Installing log jams can initiate the process of debris recruitment.
- *Riparian Planting*: Earth-moving projects and the movement of flood-control structures can disrupt the riparian ecosystem directly adjacent to rivers. Replanting of native riparian plants will support a natural vegetation community.
- *Management of Non-native and Invasive Species*: Non-native and invasive plants may be present on impaired floodplain areas and can inhibit the restoration of natural floodplain ecosystems. Actions can be taken to manage or remove these species.
- *Minimization of Land-Use Impacts*: Functional floodplains are often touted as being multifunctional, but users should not interfere with floodplain ecosystem functions. Paved paths, trails, structures, and so forth should be carefully planned to avoid conflict with inundation processes or habitat development (American Rivers 2016).

Conservation Easements and Land Acquisition

A functional floodplain should have a natural land cover that is conducive to inundation and a setting for ecological processes. Determining what amount of land is adequate to support floodplain functions is often challenging and location-specific. Ideally, enough floodplain area should be secured to contain the minimum floodplain-activating flow for a river–floodplain system. Conservation easements and land acquisition measures can be implemented to secure the necessary space for floodplain processes to occur while maximizing multiple co-benefits.

Conservation Easements

Conservation easements provide an alternative approach to land conservation. They involve voluntary legal contracts that allow lands to remain in private ownership yet restrict the rights of the property owners in specific ways that facilitate conservation. The landowner often receives financial compensation or tax breaks in return for this restriction. Easements can be used to maintain a property in its natural state or provide protection for natural attributes, or to limit the types of use to which the property may be put. They can apply forever (perpetual easements) or for a few years (term easements). They may be acquired by gift or by purchase, but obtaining an easement does not involve the transfer of fee title to the land. By acquiring an easement, a conservation agency acquires a right to undertake or to prohibit certain activities on property belonging to someone else. The landowner maintains the right to all uses that would not interfere with the activities specified. The terms of an easement can be drafted to meet the needs of the locality and the landowner. In the context of floodplain management, a conservation easement might stipulate a floodplain be maintained in its natural or undeveloped state.

Land Acquisition

Land acquisition is where a conservation group/agency takes full ownership of a piece of land. This is well suited to areas that are subject to a threat that can only be prevented if land is owned outright, or when land is exposed to many threats that need to be addressed as one. Acquisition of floodplains may be undertaken through purchase, donation, or land exchange, and is aimed at guiding future uses and remedying past mistakes. It is most effective when it is used in combination with other floodplain management tools or in conjunction with other community development or redevelopment programs (e.g., post-flood recovery measures to acquire and remove heavily damaged structures rather than repairing or rebuilding on the floodplain). In most cases, avoided future flood damages exceed land acquisition costs. In addition to providing complete and permanent protection, acquisition provides a range of additional benefits:

- Clearing the land of development and restoring it to predevelopment conditions can expand a floodplain's flow-carrying capacity, causing a reduction in upstream flood peaks, and increase its flood storage area, resulting in reduced downstream flood peaks.
- Land restoration can increase groundwater recharge, restore wetlands, and improve wildlife habitat.
- Reduced downstream flood peaks can lower stormwater management costs.
- Acquired floodplain lands can be used to meet other community objectives, compatible with proper floodplain management (e.g., establishment of greenways, recreational space, nature study areas, or greenbelts).
- Total or partial acquisition is the only means by which a community can guarantee public use of a piece of land. Regulations can limit land use, but they cannot ensure public access (Wright 2007; Fishburn et al. 2009; American Rivers 2016; Johnson et al. 2019).

Case 9.4 FEMA-Funded Residential Property Voluntary Flood Buyouts Program, United States

The US Federal Emergency Management Agency (FEMA) funds the Residential Property Voluntary Flood Buyouts program but does not buy houses directly from the property owners. Instead, acquisition and buyout projects are administered by state and local communities, which work together to identify areas where buyouts make the most sense. Individuals may not apply directly to the state, but a community may sponsor an application on their behalf. In a property acquisition project, the community buys a private property, acquires title to it, and clears it. By law, that property – which is now public property – must remain forever open space land. The community can use it to create public parks, wildlife refuges, and so forth, but it cannot sell it to private individuals or develop it. Property owners who want to sell their property will be given a fair price for it (with 75% coming from FEMA). A buyout involves the following steps:

1) An application for assistance is prepared by local officials with input from the community and those homeowners with destroyed or severely damaged properties.
2) The state receives and reviews the application and, if suitable, sends it to FEMA for approval. FEMA in turn reviews the application to ensure it follows the rules, is environmentally sound, and represents a cost-effective use of funds.
3) Once FEMA has given its approval, the state begins the acquisition process. The community conducts the purchase and title transfer. The buildings are removed or destroyed, and the land is cleared (Missouri Department of Public Safety 2016).

Source: Missouri Department of Public Safety, "Frequently Asked Questions for Fema-Funded Residential Property Voluntary Flood Buyouts," (2016), https://www.pacificmissouri.com/DocumentCenter/View/245/FEMA-Buyout-FAQ?bidId=

Wetland Restoration

The ability of a wetland to attenuate flooding depends on two factors: wetland storage capacity and the wetland storage–outflow relationship.

Wetland Storage

Storage of water in a wetland during precipitation events attenuates and delays downstream flood peaks. The delay in flood peak is significant at the catchment scale as the time lag in flood peak from a single tributary can reduce the overall flood peak in the whole catchment. While this may increase the duration of downstream flooding, it can also reduce flood depth. Wetland storage capacity is temporally available and is based on climatic conditions and land management. It increases in the summer months when rainfall is low and decreases in the winter when it is high. Wetland storage can be increased by management such as damming the outflow route to increase available overland water storage capacity or draining the floodplain to increases soil water storage capacity.

Wetland Storage–Outflow Relationship

The relationship between storage capacity and outflow rate determines the effectiveness of any wetland in flood attenuation. Wetlands naturally drain to downstream waterbody connections through surface flow and infiltration. Infiltrated water remains in the soil, drains to the water table, or joins a subsurface runoff route. Water is lost from wetlands through a combination of evaporation and evapotranspiration. Anthropogenic drainage of wetlands can alter their storage–outflow relationship (Williams et al. 2012).

Restoring Wetlands

To restore a wetland, it is recommended that a set of steps be undertaken as listed in Table 9.2. However, each wetland is unique, so not every step will be necessary in every case (Greater Wellington Regional Council 2009).

Table 9.2 Steps to restore wetlands.

Step	Description
1) Look, learn, and plan	No two wetlands are alike. How they look and the plants and animals they contain will vary with local conditions (e.g., soil, climate, water flow). A site plan should be developed based on the following questions: • What is there now? • What was there originally? • What is the future vision for this wetland? • What are its aims? Stormwater control? Erosion control? • What will be the impacts of activities on upstream and downstream properties?
2) Investigate the water supply	The water level in the wetland and how much it fluctuates will determine the plants and animals it can support. A few variables can damage a wetland's natural cycle of flooding and drying. Damage to a wetland can occur in two main places: 1) At the "wetland" level, the cycle can be affected by drainage, including the construction of drainage ditches and filling and levelling of low-lying areas 2) At the "catchment" level – the source of the wetland's water – the cycle can be affected by the occurrence of fewer floods than average, overuse of water in streams and groundwater, or the drainage of nearby wetlands
3) Keep stock out	Fencing stock out will encourage plants to regenerate from natural seed sources, prevent stock getting trapped, and lower erosion
4) Control weeds and animal pests	Weeds are a threat to wetlands, and an audit should be undertaken to locate and identify weed infestations. Animal pests pose a significant risk to wetland bird and plant life, eating chicks and adult birds and wetland plants
5) Plant	When planting, wetlands can be divided into three zones: 1) Moist soils that flood infrequently 2) Wet soils, with temporary flooding 3) Standing water/water margin It is essential to identify desirable plants for each zone
6) Monitor progress	An ongoing program of weed and pest control should be maintained, along with a photographic record of progress

Case 9.5 Amsterdam Wetlands Project

The delta region of Laag-Holland is a 12 000 hectare nature reserve. However, it is fragmented and managed by various site managers. Landschap Noord-Holland, together with Staatsbosbeheer, Natuurmonumenten, and Recreatie Noord-Holland, is aiming to develop it into one large contiguous nature area with well-coordinated water and nature management and recreational opportunities. The Amsterdam Wetlands Project, in addition to enhancing the storage capacity of the wetlands, aims to make natural areas accessible and livable (wherever possible), maintain a healthy peat landscape with regional identity, reduce subsidence and carbon dioxide emissions, and protect unique habitats and stop the loss of biodiversity (Landschap Noord-Holland n.d.).

Source: Based on Landschap Noord-Holland, "Amsterdam Wetlands: 12,000 Hectares of Top Nature," https://www.landschapnoordholland.nl/projecten/amsterdam-wetlands.

Constructed Wetlands

Constructed wetlands are engineered systems that are designed and constructed to utilize the natural functions of wetland vegetation and soils and their microbial populations to treat contaminants in surface water, groundwater, and waste streams. They can also be used to temporarily store stormwater. There are a variety of design factors to consider in their construction:

- Generally, constructed wetlands represent about 1–2% of the total catchment area.
- Their size and design are based on the type, amount, and form of pollutant inputs.
- Mosquitoes need to be managed (Government of Western Australia Department of Water 2011; Almuktar et al. 2018).

Case 9.6 South Los Angeles Wetlands Park

The South Los Angeles Wetlands Park in the United States treats up to 14 000 gallons of stormwater runoff from the 525 acre watershed per day. Located on the site of a former bus yard and brownfield at the center of a densely populated community, this constructed wetland has become a functional and attractive part of the landscape. Stormwater runoff is collected in the northwest corner of the site, diverted to an underground pretreatment system, and circulated through the constructed wetland. Any excess water, now treated, is delivered back to the traditional stormwater system at the western side of the site. A pretreatment hydrodynamic separator removes sediment, trash, oil, grease, gasoline, and heavy metals before urban runoff enters the natural system. The wetland pools are designed to maintain a depth greater than 5 feet, helping to encourage water flow, enhance wind-driven oxygenation, limit colonization of emergent vegetation, encourage particulate matter to settle, and discourage mosquito reproduction. Exposure to UV light from the sun helps reduce the bacteria in the stormwater as it moves through the system. Other features of the park include:

- Solar-powered LED lights that provide overhead illumination on trails and in the parking lots.
- Observation decks.

- Picnic areas.
- A natural rock garden seating area.
- Educational signage (Landscape Performance Series 2011).

Source: Based on Landscape Performance Series, "South Los Angeles Wetland Park," https://www.landscapeperformance.org/case-study-briefs/south-la-wetland-park#/overview.

Green Corridors

Green corridors are natural routes for the movement of people, animals, seeds, and water. They serve multiple functions, including habitat, daily and seasonal movement, and dispersal capabilities. Linking trees, parks, green spaces, and drainage areas with vegetated pedestrian walkways, cycle paths, and transport routes increases biodiversity and enables wildlife to flourish, while building resilience to floods, droughts, heatwaves, and climate change (Aziz and Rasidi 2014; ADB 2016). Green corridors provide a wide range of benefits, as summarized in Table 9.3.

Table 9.3 Benefits of green corridors.

Benefit	Description
Ecologic	Foster high-quality habitats and ecosystems, particularly natural corridors such as riparian buffers, waterways, and aquatic habitats
	Prevent soil erosion and absorb rainwater, thereby improving drainage
	Protect against the urban heat island effect
Social	Provide recreation uses: places to play, meditate, gather, and rest
	Offer a sense of place, identity, and belonging
	Promote cultural growth
	Increase the sense of community
Environmental	Provide high quality of life by integrating nature into the urban environment
	Produce a variety of routes and greenways that help increase accessibility
	Form or transform the city by delimiting its land use
	Provide sustainable transportation

Case 9.7 ESSEN.New Ways to the Water

Between 2005 and 2013, the City of Essen in Germany, as part of its action plan, "ESSEN.New Ways to the Water," developed a 148 kilometer-long green main route network with cycling and hiking paths, linking together areas of greenery and water. Over 500 individual measures were undertaken to decouple existing developments from the mixed sewer system in order to demonstrate the multifunctionality of decentralized rainwater management systems for the improvement of water balance, flood protection during heavy rainfall, decentralized flood protection, and the urban climate. This project enhanced existing networks of green connections and made a significant contribution to ensuring that suburban spaces were connected to municipal green areas. The result is that nearly all residents of Essen (>99%) can access a green public area within 300 meters of their home (European Green Capital Award 2017).

Source: European Green Capital Award, "Essen 2017 Application," https://ec.europa.eu/environment/europeangreencapital/winning-cities/2017-essen/essen-2017-application/

References

ADB. (2016). *Nature-Based Solutions for Building Resilience in Towns and Cities: Case Studies from the Greater Mekong Subregion*. Metro Manila: Asian Development Bank.

Almuktar, S.A.A.N., S.N. Abed, and M. Scholz. (2018). "Wetlands for Wastewater Treatment and Subsequent Recycling of Treated Effluent: A Review." *Environmental Science and Pollution Research International* 25, no. 24: 23 595–23 623.

American Rivers. (2016). "Reconnecting Rivers to Floodplains: Returning Natural Functions to Restore Rivers and Benefit Communities." https://s3.amazonaws.com/american-rivers-website/wp-content/uploads/2016/06/17194413/ReconnectingFloodplains_WP_Final.pdf (accessed October 9, 2020).

Auckland Council. (n.d.). "Eligibility for the Waterway Protection Fund (WPF)." https://www.aucklandcouncil.govt.nz/grants-community-support-housing/grants/regional-grants/Pages/eligibility-for-waterway-protection-fund.aspx (accessed October 9, 2020).

Aziz, H.A. and M.H. Rasidi. (2014). "The Role of Green Corridors for Wildlife Conservation in Urban Landscape: A Literature Review." *IOP Conference Series: Earth and Environmental Science* 18: 012093.

Blackwell, M.S.A., E. Maltby, A.L. Gerritsen, and European Commission. (2006). *Ecoflood Guidelines: How to Use Floodplains for Flood Risk Reduction*. Luxembourg: Office for Official Publications of the European Communities.

City of London. (n.d.). "Stormwater Management Ponds." https://www.london.ca/residents/Sewers-Flooding/stormwater/Pages/Living-Near-Stormwater-Management-Ponds.aspx (accessed October 9, 2020).

ESPON. (2018). "Territorial Potentials for Green Infrastructure." https://www.espon.eu/sites/default/files/attachments/ESPON%20Working%20Paper%20GI.pdf (accessed October 9, 2020).

European Green Capital Award. (2017). "Essen 2017 Application." https://ec.europa.eu/environment/europeangreencapital/winning-cities/2017-essen/essen-2017-application/ (accessed October 9, 2020).

FEMA. (2017). "Floodplain and Stream Restoration." https://www.fema.gov/media-library-data/1487161136815-ecad1c0312eda2111ffa28735a4d06ad/FSR_Fact_Sheet_Feb2017_COMPLIANT.pdf (accessed October 9, 2020).

Fishburn, I.S., P. Kareiva, K.J. Gaston, and P.R. Armsworth. (2009). "The Growth of Easements as a Conservation Tool." *PLOS ONE* 4, no. 3: e4996.

Golden, H.E., and N. Hoghooghi. (2018). "Green Infrastructure and Its Catchment-Scale Effects: An Emerging Science." *WIREs Water* 5, no. 1: e1254.

Government of Western Australia Department of Water. (2011). "Constructed Wetlands for Stormwater Management." https://www.water.wa.gov.au/__data/assets/pdf_file/0004/1696/99304.pdf (accessed October 9, 2020).

Greater Wellington Regional Council. (2009). "A Beginner's Guide to Wetland Restoration." http://www.gw.govt.nz/assets/Land-Management/A-Beginners-guide-to-wetland-restoration-2009.pdf (accessed October 9, 2020).

Gunnell, K., M. Mulligan, R.A. Francis, and D.G. Hole. (2019). "Evaluating Natural Infrastructure for Flood Management within the Watersheds of Selected Global Cities." *Science of The Total Environment* 670: 411–424.

Johnson, K.A., O.E.J. Wing, P.D. Bates, J. Fargione, T. Kroeger, W.D. Larson, et al. (2019). "A Benefit–Cost Analysis of Floodplain Land Acquisition for US Flood Damage Reduction." *Nature Sustainability* 3: 56–62.

Kiedrzyńska, E., M. Kiedrzyński, and M. Zalewski. (2015). "Sustainable Floodplain Management for Flood Prevention and Water Quality Improvement." *Natural Hazards* 76, no. 2: 955–977.

Kiss, T., J. Nagy, I. Fehérváry, and C. Vaszkó. (2019). "(Mis) Management of Floodplain Vegetation: The Effect of Invasive Species on Vegetation Roughness and Flood Levels." *Science of The Total Environment* 686: 931–945.

Landscape Performance Series. (2011). "South Los Angeles Wetland Park." https://www.landscapeperformance.org/case-study-briefs/south-la-wetland-park#/overview (accessed October 9, 2020).

Landschap Noord-Holland. (n.d.). "Amsterdam Wetlands: 12 000 Hectares of Top Nature." https://www.landschapnoordholland.nl/projecten/amsterdam-wetlands (accessed October 9, 2020).

Miles, B., and L.E. Band. (2015). "Green Infrastructure Stormwater Management at the Watershed Scale: Urban Variable Source Area and Watershed Capacitance." *Hydrological Processes* 29, no. 9: 2268–2274.

Missouri Department of Public Safety. (2016). "Frequently Asked Questions for FEMA-Funded Residential Property Voluntary Flood Buyouts." https://www.pacificmissouri.com/DocumentCenter/View/245/FEMA-Buyout-FAQ?bidId= (accessed October 9, 2020).

National Trust. (n.d.). "River Restoration Work in the Porlock Vale." https://www.nationaltrust.org.uk/bossington/features/river-restoration-work-in-the-porlock-vale (accessed October 9, 2020).

Opperman, J.J. (2014). "A Flood of Benefits: Using Green Infrastructure to Reduce Flood Risks." https://www.conservationgateway.org/ConservationPractices/Freshwater/HabitatProtectionandRestoration/Documents/A%20Flood%20of%20Benefits%20-%20J.Opperman%20-%20May%202014.pdf (accessed October 9, 2020).

Roca, M., M. Escarameia, O. Gimeno, L. de Vilder, J.D. Simm, B. Horton, and C. Thorne. (2017). "Green Approaches in River Engineering – Supporting Implementation of Green Infrastructure." https://eprints.hrwallingford.com/1250/ (accessed October 9, 2020).

University of Arkansas. (2009). "Riparian Buffers: Functions and Values." https://www.uaex.edu/publications/PDF/FSA-5026.pdf (accessed October 9, 2020).

University of Kentucky. (2010). "Planting a Riparian Buffer." http://www2.ca.uky.edu/agcomm/pubs/id/id185/id185.pdf (accessed October 9, 2020).

Urban Drainage and Flood Control District. (2010). "USDCM Volume 3: Stormwater Quality." https://udfcd.org/wp-content/uploads/uploads/vol3%20criteria%20manual/USDCM%20 Volume%203.pdf (accessed October 9, 2020).

Williams, L., S. Harrison, and A.M. O'Hagan. (2012). "The Use of Wetlands for Flood Attenuation." https://www.antaisce.org/sites/antaisce.org/files/final_wetland_flood_ attenuation_report_2012.pdf (accessed October 9, 2020).

Wright, J. (2007). "Floodplain Management Principles and Current Practices." https://training. fema.gov/hiedu/aemrc/courses/coursetreat/fm.aspx (accessed October 9, 2020).

10

Agriculture and Forestry

Introduction

Natural water retention measures (NWRM) are a form of green infrastructure that can play an important role in catchment-scale flood risk management. At that scale, NWRM in upstream areas can help reduce the risk of downstream flooding by enhancing or restoring natural hydrological processes (including interception, evapotranspiration, infiltration, and ponding), as well as improve water quality (Berry et al. 2017; Collentine and Futter 2018). This chapter will discuss the use of NWRM on agricultural land and in forests.

Agriculture

The utilization of land for agriculture is not traditionally acknowledged as a form of retention measure. Nonetheless, through various NWRM, retention of water within a catchment can be increased to reduce flooding risk while at the same time enhancing the natural functioning of the catchment and improving agricultural productivity (GWP and WMO 2015; Collentine and Futter 2018).

NWRM can increase organic matter in the soil, promote the liquidation of the plough layer (the sealed layer created as a result of ploughing), and improve the structure of heavy soils, all of which increases the water retention capacity of the soil. Even a small increase in water retention capacity can cause the retention of a significant amount of water. For example, an increase of water retention capacity by 10 mm (equal to 10 liters per square meter) can increase retention of 100 000 cubic meters per hectare, limiting the probability of flood occurrence (GWP and WMO 2015; Amanullah et al. 2019).

NWRM can improve soil structure, reducing the loss of valuable topsoil and increasing productivity, for example, by increasing the amount of oxygen reaching crop roots. Measures that reduce runoff and raise the water table can also lessen the impacts of droughts and reduce the number of agricultural chemicals reaching waterways. Meanwhile, NWRM can prevent livestock accessing waterways, reducing the risk of them acquiring waterborne diseases (Iacob et al. 2012; SEPA 2016). The following NWRM can be implemented on agricultural lands to reduce the risk of flooding and improve water quality, while at the same time achieving a variety of co-benefits (Table 10.1) (Rosa-Schleich et al. 2019).

Regional Water Security, First Edition. Robert C. Brears.

Table 10.1 Co-benefits of implementing natural water retention measures.

Benefit Type	Examples
Ecological	Higher biodiversity Improved pest control Higher weed control and soil health Reduced soil erosion Improved nutrient management Higher carbon sequestration
Economic	Long-term yield improvement Reduced fertilizer and machinery input Higher profitability and less risk Fewer machinery costs Lower energy costs

Source: Based on Julia Rosa-Schleich et al., "Ecological-Economic Trade-Offs of Diversified Farming Systems – a Review," Ecological Economics 160 (2019).

Meadows

Meadows are areas or fields whose primary vegetation is grass or other non-woody plants used for mowing and haying. Due to their rooted soils and permanent cover, meadows provide ideal conditions for the uptake and storage of water during temporary floods (NWRM 2015d). Croplands can be restored to meadows as they offer a ready-to-seed seedbed and the ground is mostly bare, with minimal weeds and no existing native species that need to be preserved. Meadow restoration typically involves the following steps:

- *Site Assessment*: Identify the site characteristics and define the goals for the restoration. Considerations include:
  - Has the site had herbicide treatment that would prohibit seed from germinating?
  - Is there a risk of herbicide drift from neighboring crop fields?
  - Are the soils dry, moderate, or wet?
  - How long is the soil saturated in the spring?
  - Has the site been drained with drainage tiles or ditches?
  - Is there a substantial build-up of sediments on-site that may require excavation?
  - Are there steep slopes that may be vulnerable to erosion?
- *Vegetation Removal*: Remove existing weeds and undesired vegetation from the site to prevent aggressive weedy species from out-competing native meadow plants. Considerations include:
  - Vegetation removal is not necessary on annual crop fields, provided seed is planted in the winter following harvest. If the planting is planned for the spring following harvest, a round of herbicide is usually necessary to control annual weeds.
- *Seedbed Preparation*: Prepare a seedbed to ensure good seed–soil contact and promote germination of planted seeds.

- *Seeding/Planting*: Select seed mixes and seeding methods that are well suited to the site and project goals. Considerations include:
  - Crop fields require minimal seedbed preparation unless crop residue is heavy enough to interfere with seeding.
- *Hydrologic Restoration*: Remove drainage features to restore the site's original soil moisture and seasonal flooding patterns.
- *Long-Term Management*: Maintain the health and diversity of the meadow into the future (The Nature Conservancy 2017).

Case 10.1 Floodplain Meadows Restoration Project in Riedstadt, Germany

In 2019, the Floodplain Meadows Restoration Project in Riedstadt was labelled the "Official Project of the United Nations Decade on Biodiversity." The project area is situated in the Holocene floodplain of the northern Upper Rhine, in Riedstadt, 40 kilometers southwest of Frankfurt. The area comprises 60 hectares of restoration fields, lying along both sides of the main dyke of the Rhine, and includes areas in the functional and ancient floodplain. Of the 60 hectares, 21 were purchased from federal subsidies, 24 were contributed by the municipality of Riedstadt, and 15 were contributed by the federal state of Hesse. The restoration of the floodplain meadows involved mainly diaspore-containing plant material from existing species-rich floodplain meadows in nature reserves (donor sites) nearby (the farthest was 10 kilometers away). The donor sites were cut, and the fresh green hay was transferred on to the restoration fields (donee sites). The transfer took place between September and mid-November, when most of the target species bear ripe seeds. Due to the shortage of donor sites, the distribution of plant material was limited to 10–20% of the donee sites, usually delivered in strips of 5–10 meters in width over the whole length. To establish agricultural use of the restored meadows, local farmers were included from the beginning. An initial roadshow with farmers and farming representatives was arranged to provide information about the project. This was followed by additional meetings and excursions with interested farmers and representatives. All the operations during the project (cutting, loading, transport, and distribution of plant material in the meadows) were conducted by local farmers, which helped increase the acceptance of the project among this group. In addition to involving farmers throughout the project's development, a public relations campaign was undertaken to inform the general public on why and how the project was undertaken. The emphasis was placed on the beauty of the meadows, their flowering aspects, and their contribution to the traditional cultural landscape. The public relations campaign involved a variety of communication tools:

- Mass media (press, specialized press, radio, TV).
- Internet.
- Excursions, guided tours.
- Lectures, conferences, congresses.
- Cooperation with educational establishments (e.g., adult education center) (City of Riedstadt 2020).

Source: Based on City of Riedstadt, "Restoration of Floodplain Meadows Along the Hessian Upper Rhine".

Buffer Strips

Buffer strips are areas of natural vegetation cover, including grass, bushes, and trees, at the margins of fields, arable land, and watercourses. Because of their permanent vegetation, they can slow runoff and absorb excess water. They also reduce the volume of suspended solids, nitrates, and phosphates originating from agricultural runoff. A summary of the multiple benefits buffer strips provide is given in Table 10.2 (Bentrup 2008; NWRM 2015a).

Table 10.2 Benefits of buffer strips.

Benefit	Description
Protection and safety	Enhances habitat for predators of pests Reduces flood water levels and erosion Reduces hazards
Water quality	Slows water runoff and enhances infiltration Traps pollutants in surface runoff Stabilizes soil Reduces bank erosion
Biodiversity	Increases habitat area Protects sensitive habitat Restores connectivity Increases access to resources Shades stream to maintain temperature
Productive soils	Reduces water runoff energy Reduces wind energy Stabilizes soil Improves soil quality Removes air pollutants
Economic opportunities	Produces marketable products Reduces energy consumption Increases property values Provides ecosystem services
Esthetic and visual quality	Enhances visual interest Screens undesirable views Filters air pollutants Separates human activities
Outdoor recreation	Increases natural areas Protects natural areas Protects soil and plant resources Provides a corridor for movement Enhances recreational experience

Sources: Natural Water Retention Measures, "Buffer Strips and Hedges," (2015); G. Bentrup, "Conservation Buffers: Design Guidelines for Buffers, Corridors, and Greenways," (2008).

Crop Rotation

Crop rotation is the alternation of subsistence, cash, and cover crops with different characteristics, cultivated on the same field during successive years and following an established sequence. The principle objective is to achieve production that is profitable and sustainable while maintaining soil fertility and health. Crop rotation helps maintain or increase soil organic matter content (FAO 2010). This can reduce erosion and increase infiltration capacity, reducing downstream flood risk (NWRM 2015b). Additional benefits include a reduction in the need for fertilizers due to improved soil structure and a reduction in pesticide usage as a result of lower pest populations (Boyabatlı et al. 2019). Table 10.3 lists the principles that should be taken into account when conducting crop rotation (SARE 2010).

Table 10.3 Crop rotation principles.

Principles	Description
Nitrogen supply	A legume forage crop can be followed by a high nitrogen-demanding crop to take advantage of the nitrogen supply
Nitrogen-demanding crops	Less nitrogen-demanding crops should be grown in the second or third year after a legume sod
Annual crops	The same annual crop should be grown for only one year, if possible, to decrease the likelihood of insects, diseases, and nematodes becoming a problem
Crop selection	A crop should not be followed by a closely related species, as insects, disease, and nematodes are frequently shared among them
Nematode avoidance	If specific nematodes are known problems, non-host plants can be planted for a few years to decrease populations before a very susceptible crop is planted
Healthier crops	Crop sequences should be used that promote healthier crops. Some crops do particularly well when following a particular species
Weed control	Crop sequences should be used that aid in controlling weeds
Erosion control	Longer periods of perennial crops should be applied on sloping land and highly erosive soils
Root selection	A deep-rooted crop should be grown as part of the rotation. This will scavenge the subsoil for nutrients and water, and channels left from decayed roots can promote water infiltration
Residue	Some crops should be grown that leave a significant amount of residue to help maintain organic matter levels
Grouping crops	When growing a mix of crops, they should be grouped together into blocks according to plant family, timing, type of crop, or cultural practices (e.g., irrigated, plastic mulch)
Water availability	In regions with limited rainfall and irrigation, high water-use crops may not leave enough moisture in the soil for the next crop in the rotation
Flexibility	It is important to be adaptable to annual climate change crop price variations and the development of soil pathogens and plant-parasitic nematodes

Source: Based on SARE, "General Principles".

Conservation Tillage

Tillage includes all operations of seedbed preparation that optimize soil and environmental conditions for seed germination, seeding establishment, and crop growth. Repetitive tillage degrades the soil structure and its potential to hold moisture, reduces the amount of organic matter in the soil, breaks up aggregates, and reduces the population of soil fauna (e.g., earthworms) that contribute to nutrient cycling and soil structure (FAO 2005). In contrast, conservation tillage excludes conventional tillage operations that invert the soil and bury crop residues. It slows water movement, which reduces the amount of soil erosion, potentially increasing infiltration. There are five types of conservation tillage systems:

1) *No-Tillage (Slot Planting)*: This is a one-pass planting and fertilizer operation in which the soil and the surface residues are minimally disturbed. No-tillage systems eliminate all pre-planting mechanical seedbed preparation except for the opening of a narrow strip or small hole in the ground for seed placement to ensure adequate seed–soil contact. The entire soil surface is covered by crop residue mulch or killed sod.
2) *Mulch Tillage*: Mulch tillage techniques focus on causing the least possible soil disturbance and leaving the maximum crop residue on the soil surface, while at the same time obtaining quick germination, adequate stand, and a satisfactory yield.
3) *Strip or Zonal Tillage*: The seedbed is divided into a seedling zone and a soil management zone. The seedling zone is mechanically tilled to optimize the soil and microclimate environment for germination and seeding establishment. The interrow zone is left undisturbed and protected by mulch.
4) *Ridge Till (Including No-Till on Ridges)*: The soil is left undisturbed before planting, but about one-third of its surface is tilled at planting with sweeps or row cleaners. Planting of row crops is done on preformed cultivated ridges, while weeds are controlled by herbicides.
5) *Low-Till*: Low-till agriculture consists of a combination of a crop harvest that leaves at least 30% of crop residue on the soil surface during the critical soil erosion period and some surface work (low-till). This slows water movement, which reduces the amount of soil erosion and potentially leads to greater infiltration (NWRM 2015c; Williams et al. 2016; Eurostat 2019).

Cover Crops

Cover crops are plants that are grown to provide soil cover and improve the physical, chemical, and biological characteristics of the soil. This is one of the best practices for improving organic matter levels, and thus soil quality. A variety of typical cover crops are listed in Table 10.4. The benefits of their growth include:

- Increasing soil water storage.
- Preventing erosion by anchoring soil and lessening the impact of raindrops.
- Adding plant material to the soil for organic matter replenishment.
- Binding excess nutrients in the soil and preventing leaching.

Table 10.4 Common cover crops.

Crop		Description
Legumes: Sown when "free" nitrogen is required. Generally need good drainage and fertility. Often sown with a nurse crop such as oats, or in mixes with perennial grasses	Red clover	A short-lived perennial that is somewhat tolerant of acid or poorly drained soils
	White clover	A low-growing perennial, tolerant of shade and slightly acid soil
	Sweet clover	A biennial that is deep-rooted and adapted to a wide range of soils
	Hairy vetch	An increasingly popular cover crop. Can fix large amounts of nitrogen
	Alfalfa	Requires deep, well-drained soil with a pH near neutral for good growth
Non-legumes: Selected when nitrogen contribution to the soil is not a priority	Winter rye	A typical winter cover crop. Hardy and adapted to a wide range of conditions, with an inexpensive seed
	Oat	Used as a winter cover crop to protect soil without requiring intensive management in the spring due to being frost-killed
	Ryegrass	A low-growing cover crop that produces an extensive root system which is good at capturing leftover nitrogen
	Sudangrass and sorghum-sudangrass	A fast-growing, warm-season crop that requires good fertility and moisture to perform well
	Buckwheat	A fast-growing summer annual that can be used to protect the soil and suppress weeds for a month or two between spring and fall cash crops. Grows well on acid and low-phosphorous soils
	Annual field brome	A winter annual grass. Establishes rapidly and has an extensive fibrous root system that contributes organic matter to the soil
	Japanese millet	A summer annual grass. Fast-growing and competes well with weeds
Mixtures	Legumes and grasses are often mixed as cover crops to hedge against failure of one and to get some of the benefits of both. The grass will usually establish quickly, holding soil in place and "nursing" the legume along. By taking available soil nitrogen, the grass promotes nitrogen fixation by the legume	

- Fixing nitrogen in the soil for future use.
- Providing habitat for beneficial insects and other organisms.
- Moderating soil temperatures and thereby protecting soil organisms (FAO 2010; Basche et al. 2016; UMassExtension n.d.).

Mulch

Mulches are materials placed on soil to protect it against raindrop impact and erosion. Crop residue mulching is a system of maintaining a protective cover of vegetative residues on the soil surface. It is useful when satisfactory plant cover cannot be established rapidly enough to offset erosion. Mulching adds organic matter to the soil, reduces weed growth, and reduces erosion during the period when the ground is covered in the mulch. There are two main mulching systems:

- *In Situ*: Plant residues remain where they fall on the ground.
- *Cut-and-Carry*: Plant residues are brought from elsewhere and used as mulch (FAO 2005).

In addition to reducing erosion, mulching provides numerous benefits, including:

- Enhanced water availability to crops due to better infiltration into the soil and less evaporation from it.
- Weed control.
- Less extreme changes in soil temperature.
- Reduced splashing of soil on to leaves, fruits, and vegetables (reducing diseases).
- Reduced infestations of certain pests.

Case 10.2 New York State Helping Farmers Protect Soil and Water Quality

The New York State Soil and Water Conservation Committee works to advance on-farm, comprehensive natural resources management through the support of local Soil and Water Conservation Districts. The Soil and Water Committee operates under the leadership of the Department of Agriculture and Markets to help farmers address environmental challenges and protect New York's natural resources. It provides funding through a variety of programs, including the Climate Resilient Farming Program, which aims to reduce the impact of agriculture on climate change (mitigation) and increase the resilience of New York State farms in the face of a changing climate (adaptation). Launched in 2015, it has so far rewarded $8 million to 121 farms to implement best management practice systems that reduce greenhouse gas emissions, increase carbon storage in soils, and improve the health and resiliency of farms. Funds are eligible for (track 1) manure storage cover and flare systems, (track 2) water management projects, and (track 3) Healthy Soils NY to improve soil health management practice system activities (Table 10.5) (Department of Agriculture and Markets n.d.).

Table 10.5 New York Climate Resilient Farming Program funding.

Track	Eligible Activities	Funding Available and Cost Share/Match
Track 1: Agriculture waste storage cover and flare systems	Waste storage and transfer system Manure and agricultural waste treatment system Nutrient management system – cultural	$2 million Up to 75% total eligible implementation costs
Track 2: Water management projects	Riparian buffer system Stream corridor and shoreline management system Erosion control system – structural Irrigation water management system Access control system Prescribed rotational grazing system	$1 million Up to 75% total eligible implementation costs
Track 3: Healthy Soils NY (soil health management practice systems)	Soil conservation system – cultural (Note: cover crop practices will be awarded on a per-acre basis for a 3 year term) Erosion control system – structural Prescribed rotational grazing system Riparian buffer system	$1 million Up to 75% total eligible implementation costs
Total Available Funding		**$4 million**

Source: Department of Agriculture and Markets, "Rfp 0212 – Climate Resilient Farming Round 5".

Forestry

Forests regulate streamflow, support groundwater recharge, and, through evapotranspiration, contribute to cloud generation and precipitation. They can store or recycle substantial amounts of water, thereby moderating floods. Integrating forests into the landscape, particularly in heavily deforested regions, is a cost-effective solution for flood mitigation. In water-rich areas, fast-growing, high water-consuming tree species will likely reduce flood risk. In contrast, in water-limited areas, slow-growing, low water-consuming tree species can increase infiltration and help moderate flooding (Ellison et al. 2017; Asdak et al. 2018; Bhattacharjee and Behera 2018; FAO 2019).

In addition to flood mitigation, forests improve water quality by filtering and cleaning water (leaves and root systems can trap or convert harmful toxins, helping prevent impurities from entering water systems) and controlling sediment (sedimentation can cause water pollution, adversely impacting habitats, and fill up reservoirs). Furthermore, forests protect habitats with tree cover, sheltering breeding grounds for aquatic species, providing nutrients, and cooling water (Fiquepron et al. 2013; Neary et al. 2009; IUCN 2016; de Mello et al. 2018; NIWA n.d.).

Forest Riparian Buffers

Forest riparian buffers can include a mixture of trees, shrubs, and grasses, all of which provide flood control. Trees, with their deep roots, diffuse the energy of floodwaters, reducing damage downstream, and thus flood frequency and severity, as well as associated damage to life, property, and infrastructure. Forest riparian buffers also stabilize eroding stream banks, support food and habitat for aquatic life, provide shade for species, filter out pollutants such as fertilizers, chemicals, and sediment, preserve natural corridors for wildlife, and increase the recreational potential for fishing, canoeing, and hiking. It is recommended that the forest buffer width be, at a minimum, 100 feet for perennial or intermittent streams, rivers, lakes, ponds, and reservoirs. If a water body is of exceptional value, the average width should be increased to 150 feet. Forest riparian buffers that are 100 feet or more in width sustain long-term protection of aquatic resources because they contain a critical mass or sustainable width that is essential for long-term sediment and nutrient reductions (Tiwari et al. 2016; Pennsylvania DCNR 2020).

Case 10.3 Pennsylvania's Multifunctional Forest Riparian Buffers

Pennsylvania's Department of Conservation and Natural Resources (DCNR) is leading a broad initiative to increase riparian forest buffer establishment statewide. It is supporting both conventional and multifunctional riparian forest buffers. Multifunctional buffers provide opportunities for harvesting products such as nuts, berries, woody floras, forbs, and potentially woody biomass. Inputs such as fertilizers and manure are not permitted, and harvesting is not allowed in the first 15 feet from the edge of the streambank. Over 2017–2019, DCNR offered grants to eligible applicants for the installation of multifunctional riparian forest buffers throughout the state. Applicants included government agencies, nongovernmental organizations, private businesses, and Pennsylvania residents. The total amount of funding available was $3 million, awarded at $1 million per year for each of the 3 years. With no minimum or maximum grant request amount and no matching requirement, grant funds were eligible for use to purchase materials and services related to site preparation, buffer installation, and buffer establishment. Up to 5% of the grant amount was eligible to be used to cover administrative costs such as staff time dedicated to landowner outreach and education, legal or advertisement fees, and site prioritization and design (Pennsylvania DCNR n.d.).

Source: Pennsylvania Department of Conservation and Natural Resources, "Rivers Conservation Grants".

Floodplain Forests

The essential natural function of a floodplain is to store water and sediment derived from the river basin, especially during floods. Peak annual flows that cannot be accommodated within the channel spill on to the floodplain. The water thus stored on the floodplain is slowly released after the peak has passed, reducing the peak discharge. When floodplains are forested, the forest contributes to this natural flood retention function by increasing the hydraulic roughness of the floodplain area, slowing the release of water stored on the surface even more (Hughes 2003). Floodplain forests occupy the lower areas of river catchments. Their boundaries correspond with the areas disturbed by river flooding, and they generally have shallow water tables (Hughes et al. 2012).

Healthy Floodplain Forests

Traditional flood management has involved river managers reducing the uncertainties of river flows by controlling floods and inhibiting bank erosion, sedimentation, and channel movement. However, floodplain forests cannot survive in highly controlled river systems. Instead, a healthy floodplain forest requires four elements:

1) A wide variety of river flows.
2) Moist and open sediment sites for regeneration.
3) Seeds and vegetative material to propagate.
4) Seedling-friendly water tables for tree establishment.

To ensure floodplain forests receive all four essential elements, disturbances need to be managed while ensuring the floodplain experiences floods and receives all the water, sediments, and organic debris that they carry. Without these, the forest will not be able to continue functioning naturally. Similarly, if the goal is to establish a new floodplain forest as part of a river restoration project, all these features must be present, though tress might also be planted. The essential needs of healthy floodplain forests are listed in Table 10.6 (Hughes 2003).

Scales of Floodplain Forest Conservation and Restoration

The conservation and restoration of floodplain forests can take place at the catchment scale and the local scale. Catchment-scale management involves managing physical processes in one or more places in the catchment so that they eventually influence the disturbances occurring in the floodplain zone. This type of disturbance management is "indirect," but it is highly desirable for the successful, long-term, and self-sustaining restoration or management of floodplain forests. It encourages the river to flood and the channel to move and create its sites for regeneration. Nonetheless, there are some challenges to successful catchment-scale management, including:

- Difficulty in predicting where disturbances will influence the floodplain zone.
- Lack of control.
- Skepticism among river managers and riparian owners, due to the unpredictability and lack of control.
- Difficulty where there are too many human interventions in a fragmented river system.
- Need for consensus among many stakeholders.

A variety of measures can be implemented to manage disturbance at the catchment scale:

- *Managing Flows Downstream of Dams*: This involves managing the physical processes that drive the forest ecosystem. It is a flexible measure, as floodplain forests do not need the same flows every year.
- *Managing Water Abstraction Levels*: The amount of water removed from a river for human uses affects how much is left for ecological functions downstream. Over-abstraction can lead to rivers drying out during the low-water season or to rapid drops in hydrography at critical times for tree seedling establishment. Capping abstraction levels and strictly controlling the timing and rate of abstraction can help favor floodplain forest management.
- *Managing Sediment*: Sand and gravel extraction from riverbeds and floodplain zones leads to significant changes in the sediment load carried in rivers. Often, the total load decreases, and sediments become finer. The result is downcutting of river channels.

Table 10.6 Essential needs of healthy floodplain forests.

Need	Description
Water flows	Regular low to medium flows that replenish and maintain floodplain water tables. These flows allow established trees to grow Periodic high flows that cause channel movement and sediment deposition. These provide potential regeneration sites Well-timed high and low flows through the growing season to allow delivery of seeds to the floodplain and establishment of seedling Gently tapered flows after the flood peak so that water tables recede gradually to allow successful seedling establishment A lack of high flows during the second half of the growing season, as these can destroy young seedling
Regeneration sites	Open sites, as many of the pioneer tree species cannot tolerate competition Sites that are moist through the first growing season, so that seedling roots have access to water Sites near the water's edge, because these tend to be moist and catch organic debris Sites with different sediment types, to provide varied niches for regeneration
Propagation	Floodplain forests propagate by: • Seeds carried by the river and deposited during floods. Many of these will first have been carried by the wind and then dropped on the river • Seeds carried in the wind. Whereas seeds carried in the river always move from upstream areas to downstream areas, seeds carried in the wind tend to move in the direction of the prevailing wind • Vegetative material brought by a flood or produced locally as a result of flood disturbance
Water table conditions	The gradual recession of water tables following a flood Water tables accessible to the roots of seedlings through the first growing season Limited waterlogging

Source: Hughes, "The Flooded Forest: Guidance for Policy Makers and River Managers in Europe on the Restoration of Floodplain Forests". © 2003 University of Cambridge.

The sediment imbalance can be restored by activating sediment erosion in headwater zones and allowing sediment to redistribute itself, and by reducing the levels of aggregate extraction taking place. Both options can increase the range of sediment sites available for the regeneration of floodplain forests (Hughes 2003).

Local-scale management involves managing physical processes locally in the floodplain zone. It can also involve managing the landforms in the floodplain so that the physical disturbances act differently on each part. This type of disturbance management is more direct, but it only has a relatively local effect unless many local projects interact to produce catchment-scale responses. Local-scale management is easier to achieve than catchment-scale management, for a few reasons:

- It is more predictable.
- It is relatively easier to control.

- River managers are already used to managing rivers at a local scale through the engineering of flood defenses.
- It can be implemented in the chosen river reaches between areas where other human interventions dominate.
- It requires consensus among a much smaller group of stakeholders.

Table 10.7 lists a variety of measures that can be implemented to manage floodplain disturbance at the local scale (Hughes 2003).

Table 10.7 Measures to manage floodplain disturbance at the local scale.

Measure	Description
Increasing channel roughness	River channels can become wider and more varied in both their stream habitats and their channel-edge habitats through removal of artificial embankments and the return of boulders to riverbeds
Reconnecting side arms in rivers	Channelization of many rivers has isolated them from their floodplains. The result is a loss of geomorphological dynamism and reduced forest diversity. Only species that are able to cope with dry conditions remain on the isolated floodplain. Restoration projects can remove some stretches of artificial banks and reopen side arms, significantly improving floodplain forests
Storing the flood	Floodwaters can be stored on floodplains in places where they cannot cause damage. In these flood storage areas, the floodwaters can contribute to the creation of wetland habitats, including floodplain forests. Measures to store floodwaters include: • *Setting Back Flood Defenses*: Setting back of flood defenses allows the river more space to move and increases flood storage capacity. The result is that flooding will take longer to go through, but the peaks will be lower. This reinstates the flood-attenuation role that natural floodplains have always had • *Lowering Flood Defenses*: River restoration projects can simultaneously increase flood storage capacity and renaturalize channel processes, enabling the restoration of floodplain habitats • *Lowering the Floodplain*: Floodplains can be lowered to increase their flood storage capacity. In these locations, nature development will take place, including the development of floodplain forests • *Mimicking the Natural Hydrograph*: The management of floodwaters in newly created flood storage areas will be critical to the success of achieving habitat targets in these schemes: – The design of drainage dykes and under-draining systems will affect how long and to what depth water remains on the floodplain following a flood – Trees and shrubs can add friction to the floodplain surface – The hydraulic roughness of the vegetation is determined by the height, density, and diameter of individual trees, the flexibility of the vegetation, the season (leaves), the depth of flooding, and the layout of the vegetation on the floodplain

Table 10.7 (Continued)

Measure	Description
Favoring sediment deposition	When it is not possible to manage sediment at a catchment scale, sedimentation can be artificially encouraged along river channel margins in specific places. These sites can permit natural revegetation of floodplain tree species
Managing woody debris	In natural rivers, most of the woody debris comes from forests in the upper catchment and floodplain forests. Without this debris, there is less varied instream and channel-edge habitat. Log jams can be constructed to diversify habitats
Planting forests	When floodplain restoration projects require planting of trees, genetic considerations are essential. Some species grow better in wetter sites, while others are more tolerant of drier ones. Because in most restoration programs genetic composition of source trees is unknown, it is recommended that cuttings are taken from a wide geographical range to ensure genetic variation
Managing riparian vegetation	In many locations, non-native species are common, especially in places where the physical processes of the river and floodplain have ceased or are much reduced. The restoration of floodplain processes will make conditions less attractive to non-native species. In other places, grazing and browsing inhibit the regeneration of floodplain forest, so fencing is necessary to allow regeneration to take place

Source: Based on Hughes, The Flooded Forest: Guidance for Policy Makers and River Managers in Europe on the Restoration of Floodplain Forests.

Case 10.4 Dynamic LIFE Lines Danube Project 2019–2023

The Dynamic LIFE Lines Danube project is an Austrian-Slovak initiative for the restoration of floodplains along the Danube river. In both countries, a total of 25 kilometers of side arms will be increasingly connected to the Danube and thus made more dynamic. These waters are the lifelines of the floodplain forests. The main objectives of the project are:

- Ecological upgrade of over 1600 hectares of riparian forests.
- Reconnection of around 14 kilometers of river branches.
- Dynamization of 11 kilometers of river branches.
- Renaturation of almost 4 kilometers of riverbank.

Some of the expected results of the project include:

- *Substantial Improvements to Ecosystem Functions and Services*: The productivity of riverine and alluvial ecosystems, including fish and increased carbon sequestration, water retention capacity and flood protection, and river ecosystem capacity for natural degradation of organic pollution, will be restored or enhanced.
- *Reconnection of Haslau-Regelsbrunn (Austria) River Branch*: This will result in the dynamization of about 10 kilometers of river branch system, improving floodplain forests and muddy bank habitats on around 450 hectares of floodplain.

- *Renaturation of Haslau-Regelsbrunn (Austria) Riverbank*: Almost 4 kilometers of natural shores and riverbanks will be created, as well as additional tributaries parallel to the Danube.
- *Reconnection of Spittelauer Arm (Austria) River Branches*: Reconnection of the river branches will involve the reactivation and dynamization of around 4 kilometers of river branch system and calming of about 100 hectares of alluvial forest and water habitats, with prospective island creation.
- *Reconnection of River Branches within the Danube (Slovakia)*: The connectivity of 11 kilometers of river branches will be restored, mainly at Gabkovo, Bratislava, Medveov, and Kovec.
- *Restoration of Floodplain Forests*: Over 1000 hectares of floodplain forests will be restored in the Danube floodplain area in Slovakia.
- *Genetic Species Secured*: The genetic sources of native tree species will be secured to restore target habitats.
- *Development of Best Practice Methodologies*: At least three best practice methodologies (on the removal of invasive alien species, securing of genetic sources of native tree species, and restoration of native alluvial forests) will be developed (European Commission 2019).

Source: Based on European Commission, "Dynamic Life Lines Danube - Dynamic Life Lines Danube".

References

Amanullah, H., A. Jan, Z. Shah, M.J. Khan, B. Parmar, and S. Fahad. (2019). "Organic Carbon Sources and Nitrogen Management Improve Biomass of Hybrid Rice (*Oryza sativa* L.) Under Nitrogen Deficient Condition." In *Advances in Rice Research for Abiotic Stress Tolerance*, edited by M. Hasanuzzaman, M. Fujita, K. Nahar, and J.K. Biswas. Cambridge: Woodhead, pp. 447–467.

Asdak, C., S. Supian, and Subiyanto. (2018). "Watershed Management Strategies for Flood Mitigation: A Case Study of Jakarta's Flooding." *Weather and Climate Extremes* 21: 117–122.

Basche, A.D., T.C. Kaspar, S.V. Archontoulis, D.B. Jaynes, T.J. Sauer, T.B. Parkin, and F.E. Miguez. (2016). "Soil Water Improvements with the Long-Term Use of a Winter Rye Cover Crop." *Agricultural Water Management* 172: 40–50.

Bentrup, G. (2008). "Conservation Buffers: Design Guidelines for Buffers, Corridors, and Greenways." https://www.fs.usda.gov/nac/buffers/docs/conservation_buffers.pdf (accessed October 9, 2020).

Berry, P., F. Yassin, K. Belcher, and K.-E. Lindenschmidt. (2017). "An Economic Assessment of Local Farm Multi-purpose Surface Water Retention Systems in a Canadian Prairie Setting." *Applied Water Science* 7, no. 8: 4461–4478.

Bhattacharjee, K. and B. Behera. (2018). "Does Forest Cover Help Prevent Flood Damage? Empirical Evidence from India." *Global Environmental Change* 53: 78–89.

Boyabatlı, O., J. Nasiry, and Y. Zhou. (2019). "Crop Planning in Sustainable Agriculture: Dynamic Farmland Allocation in the Presence of Crop Rotation Benefits." *Management Science* 65, no. 5: 2060–2076.

City of Riedstadt. (2020). "Restoration of Floodplain Meadows Along the Hessian Upper Rhine." https://www.riedstadt.de/stromtalwiesen/english-version.html (accessed October 9, 2020).

Collentine, D. and M.N. Futter. (2018). "Realising the Potential of Natural Water Retention Measures in Catchment Flood Management: Trade-Offs and Matching Interests." *Journal of Flood Risk Management* 11, no. 1: 76–84.

De Mello, K., R.A. Valente, T.O. Randhir, and C.A. Vettorazzi. (2018). "Impacts of Tropical Forest Cover on Water Quality in Agricultural Watersheds in Southeastern Brazil." *Ecological Indicators* 93: 1293–1301.

Department of Agriculture and Markets. (n.d.). "RFP 0212 – Climate Resilient Farming Round 5." https://agriculture.ny.gov/soil-and-water/rfp-0212-climate-resilient-farming-round-5 (accessed October 9, 2020).

Ellison, D., C.E. Morris, B. Locatelli, D. Sheil, J. Cohen, D. Murdiyarso, et al. (2017). "Trees, Forests and Water: Cool Insights for a Hot World." *Global Environmental Change* 43: 51–61.

European Commission. (2019). "Dynamic LIFE Lines Danube – Dynamic LIFE Lines Danube." http://ec.europa.eu/environment/life/project/Projects/index.cfm?fuseaction=search.dspPage&n_proj_id=7218 (accessed October 9, 2020).

Eurostat. (2019). "Glossary: Conservational Tillage." https://ec.europa.eu/eurostat/statistics-explained/index.php/Glossary:Conservational_tillage (accessed October 9, 2020).

FAO. (2005). "The Importance of Soil Organic Matter: Key to Drought-Resistant Soil and Sustained Food Production." http://www.fao.org/3/a0100e/a0100e00.htm#Contents (accessed October 9, 2020).

FAO. (2010). "Green Manure/Cover Crops and Crop Rotation in Conservation Agriculture on Small Farms" http://www.fao.org/fileadmin/user_upload/agp/icm12.pdf (accessed October 9, 2020).

FAO. (2019). "Advancing the Forest and Water Nexus: A Capacity Development Facilitation Guide." http://www.fao.org/3/ca6483en/CA6483EN.pdf (accessed October 9, 2020).

Fiquepron, J., S. Garcia, and A. Stenger. (2013). "Land Use Impact on Water Quality: Valuing Forest Services in Terms of the Water Supply Sector." *Journal of Environmental Management* 126: 113–121.

GWP and WMO. (2015). "Natural Small Water Retention Measures: Combining Drought Mitigation, Flood Protection, and Biodiversity Conservation." https://www.gwp.org/globalassets/global/gwp-cee_files/idmp-cee/idmp-nswrm-final-pdf-small.pdf (accessed October 9, 2020).

Hughes, F.M.R. (2003). "The Flooded Forest: Guidance for Policy Makers and River Managers in Europe on the Restoration of Floodplain Forests." https://www.geog.cam.ac.uk/research/projects/flobar2/reports/final/flobar2.pdf (accessed October 9, 2020).

Hughes, F.M.R., M. González del Tánago, and J.O. Mountford. (2012). "Restoring Floodplain Forests in Europe." In *A Goal-Oriented Approach to Forest Landscape Restoration*, edited by J. Stanturf, P. Madsen, and D. Lamb. Dordrecht: Springer, pp. 393–422.

Iacob, O., J. Rowan, I. Brown, and C. Ellis. (2012). "Natural Flood Management as a Climate Change Adaptation Option Assessed Using an Ecosystem Services Approach." BHS Eleventh National Symposium, Hydrology for a Changing World, Dundee. doi: 10.7558/bhs.2012.ns26.

IUCN. (2016). “Infographic: How Forests Impact Water Quality and Quantity.” https://www.iucn.org/content/infographic-how-forests-impact-water-quality-and-quantity (accessed October 9, 2020).

Nature Conservancy, The. (2017). “Restoring Your Crop Field to Conservation Meadow.” https://www.nature.org/content/dam/tnc/nature/en/documents/Restoration-Guide-Crop-to-Conservation-Meadow.pdf (accessed October 9, 2020).

Neary, D.G., G.G. Ice, and C.R. Jackson. (2009). “Linkages between Forest Soils and Water Quality and Quantity.” *Forest Ecology and Management* 258, no. 10: 2269–2281.

NIWA. (n.d.). “Sediment and Forestry.” https://niwa.co.nz/our-science/freshwater/tools/kaitiaki_tools/land-use/forestry/impacts/sediment-and-forestry (accessed October 9, 2020).

NWRM. (2015a). “Buffer Strips and Hedges.” http://nwrm.eu/measure/buffer-strips-and-hedges (accessed October 9, 2020).

NWRM. (2015b). “Crop Rotation.” http://nwrm.eu/sites/default/files/nwrm_ressources/a3_-_crop_rotation_0.pdf (accessed October 9, 2020).

NWRM. (2015c). “Low Till Agriculture.” http://nwrm.eu/measure/low-till-agriculture (accessed October 9, 2020).

NWRM. (2015d). “Meadows and Pastures.” http://nwrm.eu/sites/default/files/nwrm_ressources/a1_-_meadows_and_pastures_0.pdf (accessed October 9, 2020).

Pennsylvania DCNR. (2020). “Forest Buffers Along Waterways.” https://www.dcnr.pa.gov/Conservation/Water/RiparianBuffers/Pages/default.aspx (accessed October 9, 2020).

Pennsylvania DCNR. (n.d.). “Rivers Conservation Grants.” https://www.dcnr.pa.gov/Communities/Grants/RiversConservationandRiparianBufferGrants/Pages/default.aspx (accessed October 9, 2020).

Rosa-Schleich, J., J. Loos, O. Mußhoff, and T. Tscharntke. (2019). “Ecological-Economic Trade-Offs of Diversified Farming Systems – A Review.” *Ecological Economics* 160: 251–263.

SARE. (2010). “General Principles.” https://www.sare.org/publications/building-soils-for-better-crops/crop-rotations/general-principles/ (accessed October 9, 2020).

SEPA. (2016). “Natural Flood Management Handbook.” https://www.sepa.org.uk/media/163560/sepa-natural-flood-management-handbook1.pdf (accessed October 9, 2020).

Tiwari, T., J. Lundström, L. Kuglerová, H. Laudon, K. Öhman, and A.M. Ågren. (2016). “Cost of Riparian Buffer Zones: A Comparison of Hydrologically Adapted Site-Specific Riparian Buffers with Traditional Fixed Widths.” *Water Resources Research* 52, no. 2: 1056–1069.

UMassExtension. (n.d.). “Cover Crops & Green Manures.” https://ag.umass.edu/fruit/ne-small-fruit-management-guide/general-information/cover-crops-green-manures (accessed October 9, 2020).

Williams, A., A.S. Davis, P.M. Ewing, A.S. Grandy, D.A. Kane, R.T. Koide, et al. (2016). “A Comparison of Soil Hydrothermal Properties in Zonal and Uniform Tillage Systems across the US Corn Belt.” *Geoderma* 273: 12–19.

Part III

Case Studies, Best Practices, and Conclusions

11

Case Studies of Regions Implementing Demand Management and Green Infrastructure to Achieve Regional Water Security

Introduction

This chapter provides case studies of leading locations from around the world that have implemented a variety of demand management and green infrastructure initiatives to achieve regional water security for humans while reducing environmental degradation.

Case 11.1 City West Water
City West Water is one of three Metropolitan Melbourne water businesses owned by the Victorian Government in Australia. The water utility provides drinking water, sewerage, trade waste, and recycled water services to customers in Melbourne's central business district and inner and western suburbs. City West Water has implemented a range of demand management and green infrastructure initiatives to achieve regional water security.

Water Prices

Over the past 5 years, Melbourne has experienced warmer, drier weather conditions with less rain, which has led to water storage levels dropping to around 50% of capacity. The Victorian Desalination Plant increases the resilience of the city's water supply, supplementing water from protected catchments. The Desalination Plant can meet one-third of Melbourne's water needs, and in 2019–2020, 125 billion liters of water were ordered to help storages stabilize and recover. Since July 1, 2019, all customer water usage has covered the cost of the water ordered from the Desalination Plant, with the average cross-Melbourne quarterly residential bill increasing by $2.50. The revised 2019/2020 charges for residential and nonresidential customers are listed in Table 11.1 and Table 11.2, respectively (City West Water n.d. b).

Regional Water Security, First Edition. Robert C. Brears.

Table 11.1 City West Water residential charges.

	Old Price (July 1, 2018–June 30, 2019)	New Price (July 1, 2019–June 30, 2020)	Price Change Compared to Previous Year
Water network charge	$57.79 per quarter	$57.86 per quarter	$0.07 per quarter
Water usage charge (per kiloliter)			
Block 1: First 440 liters of water used per day	$2.4904	$2.6883	$0.1979
Block 2: Water usage above 440 liters per day	$2.9312	$3.1350	$0.2038

Source: Modified from City West Water, "Charges Explained".

Table 11.2 City West Water nonresidential charges.

	Old Price (July 1, 2018–June 30, 2019)	New Price (July 1, 2019–June 30, 2020)	Price Increase Compared to Previous Year
Water network charge	$84.18 per quarter	$84.29 per quarter	$0.11 per quarter
Water usage charge	$2.5970 per kiloliter	$2.7963 per kiloliter	$0.1993 per kiloliter
Sewerage network charge	$113.26 per quarter	$113.61 per quarter	$0.35 per quarter
Sewage disposal charge	$1.7068 per kiloliter	$1.7295 per kiloliter	$0.0227 per kiloliter

Source: Modified from City West Water, "Charges Explained".

Saving Water at Home

Around 40% of all water used in the home is in the bathroom. To reduce water consumption and enhance efficiency, City West Water provides residential customers with two initiatives:

- *Toilet Replacement Program*: Residential customers can replace their toilet with one of three different 4-star Water Efficiency Labelling and Standard (WELS)-rated dual flush models:
  - Stylus Symphony Connector Suite (white) 4-star WELS, 4.5 L/3 L flush, with a price of $434.50.
  - American Standard Heron Suite (white) 4-star WELS, 4.5 L/3 L flush, with a price of $554.00.
  - American Standard Studio – Back to Wall Suite (white) 4-star WELS, 4.5 L/3 L flush, with a price of $621.00.
- *Shower Head Exchange Program*: Residential customers can swap their old shower heads for new water-efficient ones for free. Customers follow a step-by-step guide and bring their used shower head to the nearest over eight council locations (City West Water n.d. e).

Digital Water Metering Joint Program

City West Water, in cooperation with South East Water and Yarra Valley Water, is exploring digital water metering. The three utilities are working toward deciding on how and when to roll out digital meters based on customer benefits. Specifically:

- *Customer Research*: City West Water, South East Water, and Yarra Valley Water are working together to understand the view of customers concerning future water meters and the benefits they could receive.
- *Technology Trials*: As part of the program, City West Water is trialing new technologies in metering and communications and testing the market.
- *Financial Implications*: City West Water is working to understand the financial implications of digital meters.
- *Thorough Analysis*: City West Water, South East Water, and Yarra Valley Water, in consultation with experts, will thoroughly assess whether digital water meters are viable and deliver benefits to customers, the community, and utilities. Overall, upgrading to digital meters requires a strong business case and support is required from customers, the Victorian Government, and the Essential Services Commission (an independent industry regulator for the utilities).

Digital Metering Trial

City West Water is currently undertaking a digital metering trial in Richmond and Docklands, lasting through 2020. The aim is to determine whether digital water meters can significantly reduce water loss in the network by providing more timely alerts to the utility and its customers. There is no cost to customers who participate in the trial. Instead, it is helping the utility understand the benefits of digital water meters and the practicalities of a future digital water meter replacement program, including what the financial implications may be (City West Water n.d. c).

Water Compare

Working with water suppliers across Australia, City West Water has developed an online benchmarking tool – Water Compare – to allow business customers to measure their water use against other businesses and improve their consumption. Water Compare is a free service that allows businesses to use a benchmarking calculator to evaluate their water consumption in relation to similar companies from the same industry. Data has been collected from over 3000 water users from around Australia, and water efficiency benchmarks have been created for more than 40 industry sectors to date (City West Water n.d. a).

Recycled Water

City West Water provides recycled water for use around the home and garden. Specifically, recycled water can be used to water gardens, wash cars, do laundry, and flush toilets.

Recycled water is only available in estates where "purple" colored faucets and pipes have been installed. It is provided to homes through the following steps:

1) After an initial treatment process, wastewater from homes is treated at a recycled water plant and made suitable for nonpotable use (e.g., toilet flushing, irrigation, or car washing).
2) Aquifer storage and recovery is used to store excess recycled water underground in winter so that it can be used in summer months when demand is high.
3) Recycled water is piped back through the purple pipes to be reused (City West Water n.d. d).

Stormwater Harvesting Partnership Fund

The Stormwater Harvesting Partnership Fund provides funding to councils and private entities in City West Water's service area to implement local stormwater harvesting projects that assist in keeping public parks, gardens, and sports fields green, provide fit-for-purpose water for nonpotable uses, and enhance livability in the community. Stormwater harvesting solutions provide a range of benefits, including being a source of fit-for-purpose alternative water, removing waterborne pollution and volume from waterways to protect their health, reducing pollution to bay areas, improving flood resilience, improving urban landscapes, improving amenity and livability through irrigation of passive and active open space, and assisting in relieving the urban heat island effect. Any council or private entity in City West Water's business area may apply, either singly or in a group. Successful applications must meet a range of requirements, including:

- *Asset Ownership*: The stormwater harvesting assets that are created are to be shared and owned by the Applicant and City West Water.
- *Asset Operation and Maintenance*: The Applicant will operate and maintain its assets and by agreement may also choose to operate and maintain City West Water's assets.
- *Deliver Fit-for-Purpose Harvested Stormwater to Maintain Green Public Parks, Gardens, and Sports Reserves*: Projects must demonstrate benefits such as:
  - Greening public open spaces (such as parks, gardens, and sports fields) through irrigation with stormwater.
  - Harvesting stormwater which would otherwise flow to waterways and the bay (reducing nutrient loads in waterways).
  - Saving potable water by using stormwater instead (potable substitution).
- *Co-benefits*: The project should demonstrate co-benefits being provided.
- *Maximum Funding Amount*: The maximum amount of funding available for each project application is 50% of the project cost from $100 000 up to $1 000 000. The balance of the project cost is to be provided by the Applicant and/or other funders. Any in-kind contribution would be added to the monetary contribution.
- *Co-funding by the Applicant, Council, Melbourne Water, and Other Partners*: The Applicant and other funding partners of projects must at least match the funding from City West Water on a dollar for dollar basis.

Applications will be assessed based on a weighted criterion and will be analyzed using a multicriteria assessment tool. The tool will be used to evaluate the project's benefits and

Table 11.3 City West Water Stormwater Harvesting Partnership Fund criteria.

Criterion	Description
Costs	The estimated capital cost of the project The estimated operating cost of the project Cost per kiloliter of harvested stormwater
Livability benefits	Area of public parks, gardens, and sports fields irrigated with the harvested stormwater Other livability benefits, such as increased visitations and activation of the space
Environmental benefits	The volume of harvested stormwater used The volume of potable water saved The volume of nitrogen removed from waterways through stormwater treatment and irrigation The energy required to operate the scheme, pumping versus gravity, and alternative energy sources considered
Additional benefits	Projects that demonstrate a range of environmental, livability, and community benefits will receive more favorable weighting in the application assessment process

Source: Based on "Stormwater Harvesting Partnering Fund".

costs to enable prioritization of applications for funding, allowing City West Water to obtain the best stormwater harvesting outcomes and value for money. Criteria that are considered in the assessment are listed in Table 11.3 (City West Water n.d. f).

Case 11.2 New York City Department of Environmental Protection

The New York City Department of Environmental Protection (DEP) operates a network of 19 reservoirs and three controlled lakes that cover around 2000 square miles of watershed land as far as 125 miles upstate. New York City's drinking water system is the largest unfiltered water supply in the world, delivering approximately 1 billion gallons of high-quality drinking water each day to 9 million New Yorkers. DEP has implemented a range of demand management and green infrastructure initiatives to achieve regional water security.

Municipal Water Efficiency Program

The Municipal Water Efficiency Program provides funding for water demand reduction projects in city-owned facilities. Under this program, DEP has completed retrofits and upgrades in over 450 city facilities with interagency partners including the Department of Education, Department of Parks and Recreation (DPR), New York City Fire Department, and City University of New York (CUNY). Examples of results and potential savings are listed in Table 11.4 (NYC DEP 2019).

Table 11.4 Results of the DEP Municipal Water Efficiency Program.

Agency	Water Efficiency Initiatives	Savings Achieved (May 2019)/ Potential Savings (by 2022) (Million Gallons per Day)
Department of Education	Since 2013, DEP has been upgrading bathroom fixtures in Department of Education facilities. To date, 402 facilities have been retrofitted with over 34 000 new and efficient fixtures: • Older toilets that utilize 3.5 gallons per flush are being replaced with ones that use 1.6 and 1.28 gallons per flush • Urinals that use 1.0 gallons per flush are being replaced with ones that use 0.125 gallons per flush	3.31/4.71
New York City Health and Hospitals Corporation (HHC)	The HHC operates the city's public hospitals and clinics and is the largest municipal healthcare system in the United States. Harlem Hospital is undertaking a range of retrofits, including of toilets, urinals, faucet aerators, ice machines, and an industrial dishwasher. In the next few years, DEP and HHC will expand the program by retrofitting additional HHC facilities	0.07/1.22
New York City Department of Parks and Recreation (DPR)	The DPR is the steward of more than 30 000 acres of land – 14% of New York City – including more than 5000 individual properties. DEP has partnered with DPR since 2013 on water conservation projects in city parks, beginning with the retrofitting of 400 spray showers and nine recreation centers citywide. DEP will continue this partnership with water conservation initiatives in Central Park and Prospect Park	1.1/2.73

Source: Data from NYC DEP, "Water Demand Management Plan June 2019 Annual Update," (2019).

Water Conservation and Reuse Grant Pilot Program

The Water Conservation and Reuse Grant Pilot Program provides commercial, industrial, and multi-family residential property owners with incentives to install fixture retrofits and other water efficiency technologies, such as on-site water reuse systems. Customers may be eligible to participate in the program if:

- They are an institutional, multi-family residential, or commercial customer who receives a water bill directly from DEP.
- Their water and sewage account is in good standing.
- Their water meter is connected to a functioning automated meter reader (AMR).
- Their water efficiency project will conserve at least 1 million gallons of water per year, or about 2740 gallons per day.

Table 11.5 DEP Water Conservation and Reuse Grant Pilot Program reimbursement structure.

Eligible Costs	Conditions	Reimbursement Amount
Equipment	WaterSense-certified fixture replacements	Fixed maximum unit price (see Table 11.6)
	Custom retrofits	$10 per gallon per day water savings achieved, up to 100% of documented equipment costs
Facility water use survey and design services	Third-party costs required in the development of the application	Up to 20% of total equipment costs

Table 11.6 Fixed maximum prices for WaterSense-certified fixture replacements under the DEP Water Conservation and Reuse Grant Pilot Program.

Fixture	WaterSense Specification	Maximum DEP Rebate Rate
Residential toilets (Tank or Flushometer)	Less than 1.6 gallons per flush	$125
Shower heads	Less than 2.0 gallons per minute	$45
Bathroom sink aerators	Less than 1.5 gallons per minute	$3
Commercial toilets (Flushometer)	Less than 1.28 gallons per flush	$650
Urinals	Less than 0.5 gallons per flush	$600

The following costs are eligible for reimbursement under the program (see also Tables 11.5 and 11.6):

- Third-party costs for a facility water use survey.
- Third-partly costs for design services.
- Equipment costs (including materials and supplies).

Comprehensive Water Reuse Program

Buildings with successful on-site water reuse systems are eligible for the Comprehensive Water Reuse Program. The program provides a 25% water and wastewater fee discount to DEP customers who install water reuse systems that reduce their building's water consumption by at least 25% (NYC DEP n.d. d).

Automated Meter Reading

DEP's AMR systems send readings to a computerized billing system up to four times a day. The AMR technology allows customers to view and manage their water consumption on a daily, weekly, monthly, and yearly basis through their My DEP Account.

My DEP is a centralized portal for all DEP customer online activities, providing the following services:

- Water and sewer customers:
  - Track water use.
  - Pay bills online.
  - Sign up for eBilling.
  - Sign up for online leak notification.
- Plumbing professionals:
  - Apply for and receive water and sewer permits online.
  - Pay any applicable fees online (NYC DEP n.d. a).

Green Infrastructure Grant Program

DEP offers the Green Infrastructure Grant Program for private landowners wishing to design and construct green infrastructure systems. Eligible initiatives include green roofs and infiltration projects.

Green Roof Projects

Green roof projects will undergo an advanced schematic 75% and 100% design review process, with a 2 week DEP review time for each phase. Funding is determined by green roof area and soil depth (see Table 11.7). Projects below the minimum criteria (1.5″ soil depth and 3500 square foot area) are not eligible for grant funding.

Infiltration Projects

Infiltration projects will undergo an advanced schematic 75% and 100% design review process, with a 1 month review time for each phase. If accepted, DEP will provide funding for the design and construction of these projects. Accepted projects must meet the following criteria:

- Manage 1″ of rainfall from the contributing impervious surface.
- Be cost-effective and cost more than $35 000 (NYC DEP n.d. c).

Table 11.7 Green Infrastructure Grant Program reimbursement rates for green roof projects.

Soil Depth (Inches)	Rate per Square Foot of Roof Area (for Roofs of 3500–20 000 Square Feet)[a]
1.5–1.99	$10
2.0–2.99	$15
3.0–3.99	$25
4.0+	$30

[a] The reimbursement rate for a planted area over 20 000 square feet is calculated at 50% of this rate.

Bluebelt Program

Bluebelts are ecologically rich and cost-effective drainage systems that naturally handle the runoff precipitation that falls on the city's streets and sidewalks. The Bluebelt Program, initially implemented in Staten Island, preserves natural drainage corridors (including streams, ponds, and wetlands) and enhances their ability to convey, store, and filter runoff precipitation and stormwater. In addition to reducing urban flooding and improving the health of local waterways, bluebelts provide open spaces for communities and diverse habitats for wildlife.

Staten Island Bluebelt

The Staten Island Bluebelt system drains 15 watersheds clustered at the southern end of the island, in addition to the Richmond Creek Watershed. The combined area of these 16 watersheds totals around 10000 acres. The bluebelt drainage plan connects natural drainage corridors with conventional stormwater sewers for an integrated stormwater management system. Wetlands located within the watershed areas act as flood control measures. Urban wetlands are particularly valuable as they temporarily store stormwater, helping protect adjacent and downstream property owners from flood damage. DEP has an ongoing program to purchase wetland properties for inclusion in the bluebelt system. Other publicly and privately owned areas, including New York City parkland, New York State wetland preserves, Designated Open Space, and other city-owned properties, are also incorporated into the system (NYC DEP n.d. b).

Case 11.3 The State of Washington

The State of Washington's Department of Ecology manages the state's water supply to meet the needs of people and the natural environment, in partnership with Washington communities. The state has implemented a range of demand management and green infrastructure initiatives to achieve regional water security.

Source: State of Washington Department of Ecology, "Water Rights"; "Water Banks," Retrieved from: https://ecology.wa.gov/Water-Shorelines/Water-supply/Water-rights/Trust-water-rights/Water-banks.

Water Rights

In Washington State, waters belong to the public and cannot be owned by any individual or group. Instead, a person or group may be granted a right to use a volume of water, for a defined purpose, in a specific place. Before the Department of Ecology can issue a water right permit, the proposed use must meet a four-part test:

1) Water must be available both physically and legally.
2) Water must be used beneficially.
3) Water use must be in the public's interest.
4) Water use must not impair another existing use.

The State of Washington follows the doctrine of prior appropriation, which means first users have rights senior to those issued later. This is called "first in time, first in right." If a water shortage occurs, senior rights are satisfied first, and the junior right holders can be curtailed. Water users will need a water right if they:

- Plan to use any amount of surface water (from a river, stream, spring, or lake) for any purpose.
- Plan to use groundwater (from a well) for any use, except for:
  - Single or group domestic uses of less than 5000 gallons per day.
  - Industrial uses of less than 5000 gallons per day.
  - Irrigation of lawn or noncommercial garden, a half-acre or less in size.
  - Stock water (State of Washington Department of Ecology n.d. c).

Water Banks

Water banking allows the transfer of existing water rights from willing sellers to buyers who will put the water to a new use. The goal is to move water to where it is needed the most. Creating a water bank involves consultation with the Department of Ecology and a would-be banker. If the plan is feasible and supports the public's interests, a water banking agreement may be negotiated to set up expectations for the Department of Ecology and the banker. In the agreement, at least one water right is conveyed to the Department of Ecology and held in the Trust Water Rights Program. While held in trust, the banked water right remains in good standing. The water bank can then mitigate for new water users. In addition to facilitating transfers, the Department of Ecology aims to achieve one or more of the following objectives through the creation of water banks:

- Create a reliable water supply during dry years.
- Ensure a future water supply for people, farms, and fish.
- Promote water conservation (State of Washington Department of Ecology n.d. b).

Irrigation Efficiencies Grant Program

Washington's Irrigation Efficiencies Grant Program restores instream flows in rivers and streams determined not to have enough water for fish populations and other competing needs. The Washington State Conservation Commission works with conservation districts to provide financial incentives (up to 85% of total project costs) to landowners willing to install irrigation systems that save water. The water saved helps increase the streamflow in tributaries where Endangered Species Act-listed species will benefit. In most cases, an efficient irrigation system will lead to increased crop production due to reduced water stress.

The grant funds conservation district technicians who assist landowners and water purveyors by assessing potential water savings projects and helping in their design, development, and administration. The technicians write irrigation water management plans for project participants to ensure wise water management of the new practices. They also provide monitoring to ensure success. Projects funded must meet the following criteria:

- The installation of irrigation efficiency best management practices results in a "net water savings."

- The projected water savings originate from a senior valid water right that has been beneficially used and can be protected in the primary reach.
- The net water savings are trustable.
- The implemented project benefits anadromous salmonid fish species (Washington State Conservation Commission n.d. a).

Agricultural Conservation Easements

The Washington State Conservation Commission, the agency that administers the Office of Farmland Preservation, has identified agricultural conservation easements as a tool to assist in farmland preservation. These easements:

- Are voluntary agreements that give landowners the option to sell their right to develop their agricultural property to another entity, such as a conservation district or land trust.
- Are permanent restrictions on the development and subdivision of a property title that provide a plan to protect the conservation values of the land.
- Pay landowners the portion of the property value that the development right represents and retain all other property rights.

Projects that may be considered for an agricultural conservation easement must meet the following criteria:

- The parcel proposed for preservation can continue to be used for and is large enough to sustain commercial agricultural production. The land must also be in an area that possesses the necessary market, infrastructure, and agricultural support services, and the surrounding parcel sizes and land uses must support long-term commercial agricultural production.
- The conservation district has a general plan that demonstrates a long-term commitment to agricultural land conservation.
- Without preservation, the land proposed for protection is likely to be converted to non-agricultural use in the foreseeable future (Washington State Conservation Commission n.d. b).

Source Water Protection Grant

The Source Water Protection grant program provides financial assistance to the water system and local government projects that protect public drinking water sources (Group A). A grant can be used to support drinking water protection projects that reduce the risk of contamination within a source water protection area. The program provides up to \$200 000 annually to fund source water protection projects (up to \$30 000 per project). Projects may benefit either water quality or water quantity, or both. Applicants that are eligible to apply for funding are nonprofit Group A water systems, local governments, conservation districts, nonprofit organizations, educational institutions, and tribes. Projects are expected to provide long-term benefit to drinking water quantity, with eligible activities including reducing the risk of contamination of a system's drinking water source(s) and evaluating or building up resiliency in a public water supply. Overall, funded activities must contribute to better protecting one or more public water supply sources. Examples of projects that may be eligible for funding are listed in Table 11.8 (Washington State Department of Health 2019, n.d.).

Table 11.8 Source Water Protection projects eligible for funding.

Project Type	Description and Examples of Projects Funded
Policy, research, or legislative	Policy, research, or legislative projects that identify appropriate protection measures which, when implemented, would result in reduced risk to drinking water sources. Examples of funded projects include: • Improved protection area delineation, wellhead protection plans, and watershed control plans • Research to identify contaminant sources • Water quality monitoring • Hydrologic studies (not related to water rights or other water quantity regulatory requirements) • Development of local source water protection ordinances or the critical aquifer recharge area component of critical areas ordinances • Development of restoration or conservation plans intended to improve source water quality • Identification of source catchment areas for a future easement or land acquisition • Study of the conversion from a surface water source to a protected groundwater source • Evaluation of options to address known contaminants that exceed the trigger level or federally established lifetime health advisory level • Investigation of declining source quantity due to hydrologic change • Examination of agricultural practices and carrying out of education/ outreach on best management practices
Implementation activities	Activities that reduce risks within the source water contribution area, reduce contaminant concentration in the source water, or improve a system's capacity to manage its water supply. Examples of funded projects include: • Education and information sharing (brochures, workshops, and media campaigns) • Best management practices for the reduction of pollutants that directly affect a drinking water source • Proper decommissioning of high-risk abandoned or unused (private or irrigation) wells that are close to a drinking water source • Reforestation or replanting projects in sensitive or riparian areas • Installation of signs at wellhead protection areas or watershed boundaries • Decommissioning of on-site septic systems and connection of homes to existing sewer lines to address a water quality issue or sanitary control area violation • Restoration or conservation of the ecological functions of source water protection areas • Installation of structures to divert contaminated stormwater runoff affecting a source water protection area • Installation of fencing to protect sensitive riparian source water protection areas • Implementation of pollution prevention or waste reduction projects identified in an approved source water protection plan • Development of conservation easements to protect sensitive source water protection areas • Establishment of management plans for easements or lands purchased within source water protection areas • Setting up of an ecosystem services project (carbon sequestration, habitat, open space, etc.) in a watershed to fund preservation areas • Negotiation and provision of technical documents for land purchases within source water protection areas • Establishment of a water measurement program

Table 11.8 (Continued)

Project Type	Description and Examples of Projects Funded
Security measures	Security measures to reduce the risk of contamination to the source water protection area or intake/well (all such measures must protect highly sensitive source water areas, not just facilities or equipment). Examples of funded projects include: • Fencing around sanitary control areas or intakes (purchase of materials) • Gates for access points (purchase of materials) • Security systems such as alarms, signs, cameras, locks, and lights

Source: Washington State Department of Health, "Source Water Protection Local Assistance Grant Program"; "Source Water Protection Grant Guidelines," (2019).

Floodplains by Design

Floodplains by Design is a public–private partnership led by the Department of Ecology, The Nature Conservancy, and the Puget Sound Partnership. The initiative works to accelerate integrated efforts to reduce flood risks and restore habitat along Washington's major river corridors. Its goal is to increase the resilience of floodplains in order to protect local communities and the health of the environment (Floodplains by Design n.d.).

The Department of Ecology's Floodplain Management Division administers the Floodplains by Design grant program under a biennial funding cycle. Grants are awarded on a competitive basis to eligible entities for collaborative and innovative projects throughout Washington. Projects receiving grant funding must support the integration of flood hazard reduction with ecological preservation and restoration. They may also support other community needs, provided they are part of a broader strategy, such as agriculture preservation, water quality improvement, or provision of increased recreational opportunities. The grant funding specifics are:

- *Amount of Funding Available*: The amount of funding available varies as determined by the state legislature.
- *Grant Award Limit*: The project must be achievable in 2–3 years.
- *Amount of Matching Funds Required*: Funds are set at 20%, waived for economically distressed communities.
- *Eligible Organizations*: Funding is available to local and tribal governments, port districts, flood control zone districts, flood control and dyking and drainage districts, conservation districts, municipal and quasi-municipal corporations, and nonprofits that are recognized by the Internal Revenue Service as tax-exempt.

Grant applications are evaluated on the strength of a variety of elements:

- *Evidence of Flood Hazard Reduction*: How the project will reduce the magnitude or frequency of flood damage to people, structures, or infrastructure.
- *Evidence of Benefit to Natural Floodplain Function*: The ecological benefits of the project, its significance, its up/downstream effects, and its ability to address the overall need in the project area or watershed.

Table 11.9 Benefits brought by Floodplains by Design projects.

Benefit	Description
Reduction of flood risk	38 communities have been protected 9.5 miles of new state-of-the-art levees have been built to provide flood protection 500 tons of rubbish and hazardous waste have been removed from Dungeness River 700 residences have been removed from high-risk floodplain areas
Restoration of salmon habitat	Natural processes have been restored along 25 miles of river 2500 acres of floodplains have been reconnected, with planning underway for 3000 acres more
Community benefits	Public access and recreational trails have been built on eight different rivers
Protection of working lands	500 acres have been protected for long-term agricultural use Water quality improvements have been made to 1220 acres of shellfish beds

Source: Based on Floodplains by Design, "Impact".

- *Puget Sound Projects*: Where applicable, how the project contributes to the restoration and protection of Puget Sound and how the proposed activities or strategies are consistent with the Puget Sound Partnership Action Agenda.
- *Other Benefits*: What other benefits the project will provide (e.g., agricultural, water quality, recreation, public access, community/stakeholder support).
- *Projected Future Changes*: How the project accounts for expected future changes to hydrology, sediment regimes, sea-level rise, or water supply resulting from other floodplain management efforts, land-use changes, extreme weather events, or other causes.
- *Integration of Project*: Whether the project is consistent with existing flood management or habitat plans.
- *Expenditure*: Whether funds will be spent promptly. A reasonable budget appropriate to the project scope must be provided. If applicable, it must also be shown how the project leverages existing investments (match).

Since 2013, $115 million has been used to support 36 large-scale, multi-benefit projects on 13 major floodplains across the state. Some of the benefits to date are listed in Table 11.9 (State of Washington Department of Ecology 2019, n.d. a).

Case 11.4 Welsh Water

Welsh Water is the sixth largest of the 10 regulated water and sewerage companies in England and Wales. The utility is responsible for providing over 3 million people with a continuous, high-quality supply of drinking water and for managing wastewater. Welsh Water has implemented a range of demand management and green infrastructure initiatives to achieve regional water security.

Source: Based on Welsh Water, "Free Items".

Free Water Saving Gadgets

Welsh Water has partnered with Aqualogic to provide customers with free water-saving gadgets:

- *Save a Flush Bag, 1.2 Liters*: This bag is dropped into the cistern and saves around 1 liter on every flush. It is best suited to most modern cisterns larger than 6 liters.
- *Toilet Leak Detection Strip*: This leak strip is designed to show if there are any internal leaks in a toilet.
- *Winter Lagging Kit*: This DIY pack is supplied with the highest-quality insulation items for home pipes and faucets. It can be used indoors or outdoors (Welsh Water n.d. a).

Water Efficiency Audit for Small Businesses

Most companies and organizations know how much water they use, but many do not use this knowledge to reduce their usage. Welsh Water provides a simple guide to help small businesses reduce the amount of water they use, and the amount they waste. It also offers a water efficiency audit service to help small business customers identify any hidden leakage problems, which cost businesses significant amounts of money each year. The water audit will identify areas where the business can be more water-efficient, as listed in Table 11.10.

Table 11.10 Areas in which small businesses can become more water-efficient.

Question	Efficiency Areas
Where does wastage occur?	Running or dripping faucets Faulty ball valves/overflowing cisterns Constantly flushing urinals Underground leaks
How much water is used?	Toilets: 6–9 liters per flush Sinks: 6–9 liters per bowl Showers: 30–90 liters per use Dishwater: 20–40 liters per load Laundry: 60–100 liters per load Vehicles: 100 liters per vehicle using buckets, 900 liters per vehicle using a hose Garden hoses: 8–30 liters per minute (500–1800 liters per hour) Employees (full time, no canteen): 25 liters per person per day Employees (full time, with a canteen): 40 liters per person per day
How much water is lost due to leaking faucets?	One drip per second wastes 4 liters per day Drips breaking into a stream waste 90 liters per day 1.5 mm stream wastes 320 liters per day 3 mm stream wastes 985 liters per day 6 mm stream wastes 3500 liters per day
What water-saving devices can be installed?	Toilets per cistern: dual-flush devices/water displacement devices Urinals: flow control, waterless urinals Sinks: push faucets/spray faucets Hose pipes: trigger outlets

The audit steps are as follows:

1) Identify all water fittings and check for leaks:
   - Check whether faucets are drip-free.
   - Make sure no water is being used internally and check to see if the water meter is turning (if it is, there is likely to be a leak).
   - Check all water fittings to see if there are any water-saving devices fitted.
2) Check for underground leaks. Locate and shut off the internal stop faucet to isolate the supply entering the building, and check whether the water meter is turning (if it is, there is likely to be a leak).
3) Read the water meter:
   - Do this twice daily at the same times for a week.
   - Do one reading at the start and one at the end of the working day, to record all usage during work hours.
   - Record meter readings regularly (daily, weekly, or monthly), in order to identify trends in water consumption.
4) Understand water usage. Compare previous bills to check for abnormal consumption and identify any activities that use large volumes of water during this period.
5) Analyze the results:
   - Look at the times when the building is unoccupied. Is water being used? If so, this could be due to continually flushing cisterns or a leak.
   - Look at daily trends and note any differences in the amount of water being used on different days. If so, try to identify where the water is being used and if this use is necessary.
   - Look at the amounts of water being used for large-volume activities and assess whether they are reasonable.
6) Develop an action plan:
   - Take regular meter readings and produce graphs to track usage trends.
   - Regularly check for leaks.
   - Install water efficiency devices.
   - Reduce water pressure.
   - Draw a plan to map underground water services, valves, and meters and investigate any damp patches near them.
   - Fit submeters for better monitoring.
   - Use water-efficient maintenance activities.
   - Raise staff awareness and encourage employees to report leaks.
   - Prevent burst pipes in winter by lagging external pipe works, tanks, and cisterns (Welsh Water 2019).

Standard Water Efficiency Survey

Welsh Water provides a standard water efficiency survey that assesses domestic water use on business sites (e.g., toilets, faucets, and cleaning facilities). The aim is to identify opportunities to save water and to quantify them, where possible, in volumetric terms. A Welsh Water employee will review the customer's current water consumption and specific elements of the water installation at each site. The survey will include a selection of the following activities to help the customer identify products that may need to become more water-efficient:

- Making a general inspection of domestic water systems.
- Analyzing meter readings to determine current water consumption patterns.
- Identifying site submetering.
- Assessing the possible extent of water leakage both above- and belowground.
- Assessing urinal cistern control mechanisms.
- Commenting on water management procedures, if appropriate.
- Summarizing total potential water savings (Welsh Water n.d. f).

Process Audit

Welsh Water offers business customers a process audit service that assesses water consumption within production and operational processes. The audit is carried out by a Welsh Water industrial process engineer and can be scaled to reflect the needs of the customer. The critical parts of the service include:

- *Site Walkover*: An experienced engineer will carry out a detailed observation of the customer's on-site processes, identifying areas of high consumption or wastage, plus cost-saving opportunities.
- *Data Analysis*: A detailed site data analysis will be undertaken, including efficiency correlation analysis and analyses of production, water and effluent volumes, and, where appropriate, concentrations.
- *Mass Balance*: A site-wide balance will be created, exposing unidentified losses and the costs associated with each and identifying the key cost drivers for the site.
- *Cost–Benefit Analysis*: A cost–benefit analysis will identify opportunities for cost savings and show payback periods and priorities by profitability. Welsh Water will also compare any production and water consumption data with industry benchmarks to make clear and customized recommendations on how to improve water efficiency (Welsh Water n.d. c).

Telemetry

Welsh Water offers flow monitoring services to help business customers understand their consumption patterns and identify any wasteful water usage or leakage. Water monitoring equipment can be connected to the customer's water meter, which will enable Welsh Water to track the site's water usage. The water monitoring equipment records information from the water meter and transfers it into flow data. The data are then transferred to a designated website where the customer can monitor their water flow and consumption. Customers can track their water use graphically, allowing potential problems such as leaks to be identified quickly. The information can also be shared with relevant people in the business. The benefits of this program are listed in Table 11.11 (Welsh Water n.d. e).

Table 11.11 Benefits of Welsh Water's telemetry service.

Benefit	Description
Creating baseflows	Water monitoring equipment on-site allows customers to build up a picture of their water use, helping them understand their usage, identify peaks and troughs in consumption, and determine baseline flows
Efficient water use	Automated readings can help identify waste or leaks. Baseline data can optimize a customer's water use and ensure they manage their water efficiently and economically
Rapid leak detection	Alarm functionality allows customers to set alarms and receive alerts when unusual flow levels are detected
Health and safety	Telemetry reduces the need to access a chamber to read the meter, eliminating health and safety risk

Source: Based on "Telemetry", Dŵr Cymru Cyf.

RainScape

Welsh Water is developing and using a new, innovative solution called RainScape to manage the amount of surface water entering the sewer system. The utility has invested around £80 000 in this approach in order to reduce the risk of sewer flooding and pollution and increase system capacity. RainScape will support economic development and protect customers against climate change. Using international examples of green infrastructure practices, Welsh Water has created a range of RainScape solutions that can be incorporated into new developments or installed into existing sewer systems, including basins and planters, swales, porous paving, filter strips, grass channels, and geocellular storage (Welsh Water n.d. d).

Partnership Flood Alleviation Program

Welsh Water's partnership flood alleviation program provides risk management authorities and community groups within its operating area the opportunity to work with it to manage the risk from surface water flooding, in terms of both Welsh Water's assets and those of others. The utility will consider contributing to surface water flooding alleviation projects that also benefit the public sewer network or itself in the following ways:

- Reduced risk of flooding to vulnerable communities.
- Reduced peak river flows (up to 25%) in a small number of catchments.
- Reduced surface water entering Welsh Water's networks in affected catchments.
- Reduced impact of flood events on water quality.
- Enhanced reputation.

Welsh Water will contribute to solutions that encourage a partnership approach to the management of surface water flooding using natural flood management or sustainable drainage systems (SuDS). Applications must ensure their proposed project is aligned to at least one of the benefits just listed. The level of financial contribution made by Welsh Water will be based on the scale of benefits and their expected impact on the performance of the public sewer network (Welsh Water n.d. b).

References

City West Water. (n.d. a). "Benchmarking." https://www.citywestwater.com.au/business/saving_water_energy/benchmarking.aspx (accessed October 9, 2020).
City West Water. (n.d. b). "Charges Explained." https://www.citywestwater.com.au/billing/charges_explained.aspx (accessed October 9, 2020a).
City West Water. (n.d. c). "Digital Water Metering Joint Program." https://www.citywestwater.com.au/about_us/major_projects/digital_metering_program.aspx (accessed October 9, 2020b).
City West Water. (n.d. d). "Recycled Water." https://www.citywestwater.com.au/saving_water/recycled_water.aspx (accessed October 9, 2020c).
City West Water. (n.d. e). "Saving Water at Home." https://www.citywestwater.com.au/saving_water/saving_water_at_home.aspx (accessed October 9, 2020d).
City West Water. (n.d. f). "Stormwater Harvesting Partnering Fund." https://www.citywestwater.com.au/saving_water/partnering_fund.aspx (accessed October 9, 2020e).
Floodplains by Design. (n.d.). "Our Impact." http://www.floodplainsbydesign.org/impact/ (accessed October 9, 2020).
NYC DEP. (2019). "Water Demand Management Plan June 2019 Annual Update." https://www1.nyc.gov/assets/dep/downloads/pdf/water/drinking-water/water-conservation-report2019.pdf (accessed October 9, 2020).
NYC DEP. (n.d. a). "How We Bill You." https://www1.nyc.gov/site/dep/pay-my-bills/automated-meter-reading-frequently-asked-questions.page (accessed October 9, 2020).
NYC DEP. (n.d. b). "The Bluebelt Program." https://www1.nyc.gov/site/dep/water/the-bluebelt-program.page (accessed October 9, 2020).
NYC DEP. (n.d. c). "Grant Program for Private Property Owners." https://www1.nyc.gov/site/dep/water/green-infrastructure-grant-program.page (accessed October 9, 2020a).
NYC DEP. (n.d. d). "Water Conservation & Reuse Grants." https://www1.nyc.gov/site/dep/water/water-conservation-reuse-grants.page (accessed October 9, 2020b).
State of Washington Department of Ecology. (2019). "Pre-application for a 2019–2021 Floodplains by Design Project Grant." https://ecology.wa.gov/DOE/files/b3/b33d6350-b529-4975-96bc-8edd08cf5d0a.pdf (accessed October 9, 2020).
State of Washington Department of Ecology. (n.d. a). "Floodplains by Design." https://ecology.wa.gov/About-us/How-we-operate/Grants-loans/Find-a-grant-or-loan/Floodplains-by-design (accessed October 9, 2020).
State of Washington Department of Ecology. (n.d. b). "Water Banks." https://ecology.wa.gov/Water-Shorelines/Water-supply/Water-rights/Trust-water-rights/Water-banks (accessed October 9, 2020).
State of Washington Department of Ecology. (n.d. c). "Water Rights." https://ecology.wa.gov/Water-Shorelines/Water-supply/Water-rights (accessed October 9, 2020).
Washington State Conservation Commission. (n.d. a). "Irrigation Efficiencies Grant Program." https://scc.wa.gov/iegp/ (accessed October 9, 2020).
Washington State Conservation Commission. (n.d. b). "Agricultural Easements." https://www.scc.wa.gov/ofp/agricultural-easements (accessed October 9, 2020).
Washington State Department of Health. (2019). "Source Water Protection Grant Guidelines." https://www.doh.wa.gov/Portals/1/Documents/Pubs/331-552.pdf (accessed October 9, 2020).

Washington State Department of Health. (n.d.). "Source Water Protection Local Assistance Grant Program." https://www.doh.wa.gov/CommunityandEnvironment/DrinkingWater/SourceWater/LocalAssistanceGrantProgram (accessed October 9, 2020).

Welsh Water. (2019). "Water Efficiency for Small Businesses." https://www.dwrcymru.com/-/media/Files/Business/Water-Efficiency-For-Small-Business.pdf?la=en&hash=85C436CFD3EDF42EBE5EEECF85AF67772C3729A5 (accessed October 9, 2020).

Welsh Water. (n.d. a). "Free Items." https://aqualogicwc.co.uk/store/dcww/free-items/ (accessed October 9, 2020a).

Welsh Water. (n.d. b). "Improving Flood Prevention." https://www.dwrcymru.com/en/My-Wastewater/Partnership-Flood-Alleviation-Programme.aspx (accessed October 9, 2020b).

Welsh Water. (n.d. c). "Process Audit." https://www.dwrcymru.com/en/Business/Our-Services/Process-Audit.aspx (accessed October 9, 2020c).

Welsh Water. (n.d. d). "Rainscape." https://www.dwrcymru.com/en/My-Wastewater/RainScape.aspx (accessed October 9, 2020d).

Welsh Water. (n.d. e). "Telemetry." https://www.dwrcymru.com/en/Business/Our-Services/Telemetry.aspx (accessed October 9, 2020e).

Welsh Water. (n.d. f). "Water Efficiency Survey." https://www.dwrcymru.com/en/Business/Our-Services/Water-Efficiency-Survey.aspx (accessed October 9, 2020f).

12

Best Practices

Introduction

From the case studies of regions implementing a range of policy innovations that promote the application of demand management and green infrastructure to achieve regional water security, including the extended case studies listed in Chapter 11, a range of best practices have been identified in the demand management-related areas of water allocation, water augmentation, water efficiency, and water reuse and recycling, as well as in the application of green infrastructure at the building and street level and in parks and forests, water bodies, and agricultural and forestry lands. These best practices can be implemented by other regions attempting to achieve regional water security for humans while protecting and restoring the natural environment.

Water Allocation

A series of best practices have been identified for regions reforming their water allocation regimes in order to manage the risk of shortage and adjudicate between competing users.

Water Pricing

Singapore's Public Utilities Board (PUB) is investing in and upgrading its water supply system to make it more resilient to climate change. At the same time, the costs of producing and maintaining water infrastructure are increasing. To meet rising infrastructure and treatment costs and ensure prices reflect the scarcity value of water, the price of water and wastewater for domestic and nondomestic customers has been revised upward over two phases.

Subsidies

Toronto Water helps businesses reduce their water use, which in turn lowers their operating costs, enhances their public image by demonstrating a commitment to the

Regional Water Security, First Edition. Robert C. Brears.

environment, and increases their competitiveness by saving money and attracting new customers. The following rebates are offered to businesses:

- The Capacity Buyback Program offers a free water audit to help customers identify ways of reducing water use and rewards participants that implement permanent water-saving measures with a one-time cash rebate.
- The Industrial Water Rebate Program offers a discounted water rate to manufacturers in Toronto to help support economic growth and encourage water conservation.
- The Sewer Surcharge Rebate Program provides a rebate for water not discharged into the sanitary sewer system.

Water Trading

Water in the Murray–Darling Basin in Australia can be bought or sold, either permanently or temporarily. The water is traded on markets, within catchments, between catchments (when possible), and along river systems. The water markets allow trading of rights to an ongoing share of water within a system, as well as of the amount of water distributed to users.

Basin-Level Modeling and Seasonal Forecasting

New Zealand's Water Model, developed through a partnership between research institutes and regional councils, will be used to predict how climate, land-use, and other environmental changes will affect aquatic ecosystems and water resources available for use by businesses and communities.

Water Restrictions

The City of Pasadena in California prohibits water waste, restricts outdoor watering over the spring/summer period, and requires property owners to fix leaks in their sprinkler systems. Its utility provides information to customers on how to avoid wasting water, and issues written warnings for first offences. Customers who continue to violate the ordinance face fines.

Water Augmentation

A series of best practices have been identified for regions implementing water augmentation strategies in order to increase the available supply of water through active recharge or protect water recharge areas.

Conjunctive Use and Development of Surface and Groundwater

The Sacramento Regional Water Bank in California is an innovative groundwater storage program that will improve regional water supply reliability. During dry years, recovery of stored groundwater will occur through additional pumping, resulting in a withdrawal from the Water Bank. During a wet year, local water providers can store water in the Water Bank using existing facilities.

Managed Aquifer Recharge

The Groundwater Replenishment Scheme in Perth, Western Australia provides the city with a climate-independent water source. Commissioned after a trial, the project is designed to treat recycled water to a level of purity that is equal to or greater than drinking water quality before injecting it into aquifers for later abstraction for drinking water. The next stage of the project will double the size of the scheme and involve the construction of a second Advanced Water Recycling Plant, as well as new bores for recharge.

Flood Managed Aquifer Recharge

The California Department of Water Resources, in partnership with the Merced Irrigation District, is conducting a preliminary study to determine the potential for using flood water for managed aquifer recharge (MAR). The study is a proof of concept evaluating the feasibility, effectiveness, and benefits of planning and implementing a flood MAR project. Computer models are being used to estimate flood stage, flow, and floodplain inundation for rivers and creeks within the study area and to determine recharge capacities.

Source Water Protection

In Bavaria, Germany, water protection areas have three zones, with restrictions easing with distance from the well. To establish new source water protection areas, a series of steps must be followed, including a hydrogeological study, analysis of risks to the aquifer, presentation to municipal stakeholders and public administration officials, a public display, a hearing with experts, and an announcement of the permit, including all relevant map materials.

Water Efficiency

A series of best practices have been identified for regions promoting water efficiency measures in order to reduce water scarcity and maximize the benefits provided by existing water infrastructure.

Water Metering

The City of Houston, Texas' water customers can sign up for high usage alerts and receive a notification via email if their water usage is higher than usual. They can also access usage information via computer or phone, view their historical water usage, and track their usage history. Customers will receive alerts according to their set water usage threshold.

Reducing Unaccounted-for-Water

The Victorian Government in Australia has completed a pilot program using satellite scanning technology to detect water leaks underground. The pilot involved Victoria's water corporations and VicWater partnering with a private-sector company to verify whether leaks could be detected through scans from space. Once leaks were detected, field crews were dispatched to repair them.

Public Education and Awareness

San Diego County Water Authority, California provides a variety of tools to help homeowners and landscape professionals increase water efficiency countywide and make WaterSmart living a permanent way of life. The tools include a series of online videos to help guide homeowners through the process of transforming their properties into water-efficient landscapes and the local introduction of an Environmental Protection Agency-approved training program for landscape professionals.

School Education

Actsmart Schools in the Australian Capital Territory is a free program that helps schools save energy, conserve water, increase recycling, protect biodiversity, reduce greenhouse gas emissions, and integrate sustainability into the curriculum. To conserve water, the program provides a best practice guide, links water conservation with the national curriculum, provides professional learning in water efficiency and water quality for teachers, business managers, and facilities managers, and offers examples of best practices and case studies on sustainable water usage. The program also sends officers to visit schools and provides an analysis of water usage compared to benchmark levels.

Developing a Resource-Efficiency Skill Base

Singapore's PUB has developed with Singapore Polytechnic a Water Efficiency Managers Course that equips facilities managers with the knowledge and skills to conduct water audits and apply water efficiency measures. Run over 3 days, the course covers a variety of topics, including water audits, water-efficient landscaping, and various industrial water solutions. Successful participants receive a certificate of proficiency.

Water Labeling

Australia's Water Efficiency Labelling and Standards scheme helps consumers choose more water-efficient products for their home or business. The scheme regulates plumbing products and whitegoods. All products must be registered with a regulator and labeled with the correct water rating information. The water rating label allows consumers to compare the water efficiency of different products and determine a given product's rate of water consumption.

Water Efficiency in Buildings

Los Angeles, California requires all existing large, nonresidential buildings to submit benchmarking data for water and energy use annually. The aim is to diagnose energy-inefficient buildings and aid in the implementation of energy-saving features. Every 5 years, either prescriptive or performance measures are undertaken to improve overall energy efficiency. This can involve energy and water audits, retro-commissioning, and energy and water reduction targets.

Industrial Water Efficiency

Welsh Water helps businesses lower their water and energy bills while reducing carbon emissions. A Welsh Water industrial process engineer assesses the organization's water consumption within the production and operational processes on-site. The service can be scaled to reflect the needs of the customer, but the key aspects include a site walkover, data analysis, a mass balance, and a cost–benefit analysis. Welsh Water will also compare any available production and water consumption data with industry benchmarks to make clear and bespoke recommendations on how to improve water efficiency.

Agricultural Water Efficiency

The Australian Government assists irrigators to modernize their on-farm irrigation infrastructure while returning saved water to the environment. Water savings generated from the program will help secure a sustainable future for irrigation communities. Funding is available to install new or upgrade existing infrastructure equipment and technology, improve irrigated area layout or design, and upgrade or convert to surface or subsurface drip systems and overhead spray systems.

Water Reuse and Water Recycling

A series of best practices have been identified for regions implementing a range of water reuse and water recycling technologies.

Rainwater Harvesting

Austin Water in Texas offers a rebate to help customers buy equipment for capturing rainwater. The rebate is available for systems with or without pumps. Eligibility requirements for customers include keeping the system in service for a set time, agreeing to pre- and post-installation inspections by the utility, and allowing the utility to reference the project in outreach materials.

Stormwater Harvesting

The Fitzroy Gardens Stormwater Harvesting System in Melbourne, Australia captures, treats, and stores stormwater to be reused for irrigation of the park. The system is part of a wider project that includes new public open space, new depot facilities, and a visitor center. In addition to saving drinking water, the system provides a reliable alternative water source for irrigation, reduces the amount of pollutants entering waterways, and improves the city's ability to manage the impacts of climate change.

Water Recycling

Singapore's PUB is providing funding support for high-impact and innovative projects that treat and reclaim freshwater from industrial used water for process reuse. Large companies are invited to submit proposals that yield a significant reduction in water consumption through reuse. On successful completion, the project's technology may lead to full-scale implementation in similar installations within the industrial sector.

Indirect Potable Reuse

The City of Wichita Falls, Texas' Indirect Potable Reuse System provides treated water to the city's main source of potable water. It makes Wichita Falls more drought-tolerant by extending potable supplies and fully utilizing return flows from the city's Resource Recovery Facility.

Direct Potable Reuse

The City of Flagship, Arizona is evaluating future water supplies, including the offering of Class A+ reclaimed water through Direct Potable Reuse. It is planned that after reclaimed water has been treated at a wastewater treatment facility, it will be retreated at an Advanced Water Treatment Facility. The purified water will then be held in a storage tank for final quality testing before distribution.

Onsite Nonpotable Reuse Systems

New York City's Department of Environmental Protection (DEP) provides grant assistance for the construction of nonpotable water reuse systems at both the building and the district scale. The grant's overall goal is to achieve a total water demand reduction of 1 million gallons per day. Buildings with successful on-site water reuse systems are also eligible for the Comprehensive Water Reuse Program, which provides a discount on water and wastewater fees.

Environmental Flows

The Tokyo Metropolitan Government recommends recycled water for miscellaneous use in buildings. Reclaimed water may also be used for the restoration of waterways, contributing to an increase in water volume, a recuperation of the water environment, and an increase in spaces for recreation and relaxation.

Other Alternative Water Sources

Captured Condensate

The Federal Energy Management Program in the United States provides a condensate capture potential map that allows users to identify whether they live in an area that has high potential for its application.

Atmospheric Water Generation Technology

World Vision South Africa offers an innovative mobile water generation project in drought-stricken communities. Grant funding provides a mobile, diesel-powered generator that can produce high-quality water irrespective of air quality.

Renewable Energy Desalination

Masdar in the United Arab Emirates has launched a renewable energy desalination pilot program to research and develop energy-efficient, cost-competitive desalination technologies powered by renewable energy. Commercial partners operate next-generation pilot seawater desalination plants to test a range of approaches to increasing operational efficiency, including solar-powered reverse osmosis.

Fog Harvesting

A women-led nongovernmental organization in Morocco has designed and installed the world's largest operational fog water harvesting system. The project provides potable water to mainly women and children in a handful of villages, allowing them more time to work and study.

Green Buildings and Green Streets

A series of best practices have been identified for regions implementing green infrastructure at the building and street level.

Green Roofs

Rotterdam in The Netherlands provides a subsidy for the construction of green roofs. Homeowners, companies, associations, and foundations are all eligible. The amount provided is based on the roof's water storage capacity and green area. The city encourages the development of multifunctional roofs such as green roofs with solar panels, blue roofs that function as water gardens with a water buffer during extreme rainfall, and green roofs that provide a space for working, conferencing, or gardening.

Rain Barrels and Cisterns

The San Francisco Public Utilities Commission has partnered with the Urban Farmer Store to provide residents, businesses, and schools with free rain barrels. Customers can also receive a discount on cisterns.

Rain Gardens

New York City's DEP is building rain gardens and other types of green infrastructure to manage stormwater and improve water quality in the city's waterways. The building of neighborhood rain gardens follows a process that includes site selection (only locations that can effectively absorb stormwater are considered), construction (before construction, the city will reach out to surrounding property owners), and maintenance (the city is responsible for maintenance).

Bioswales

The City of Vancouver has implemented a range of green infrastructure projects, including a green street project that aims to increase access to green space, enhance rainwater management, and improve neighborhood amenities. The project includes dedicated bioretention systems, seating areas, bike racks, a drinking water fountain, a rain garden, and a bioswale that improves water quality and reduces combined sewer overflows.

Permeable Pavement

The Permeable Pavement Project in the City of Tacoma, Washington is converting failed residential streets into green infrastructure using permeable pavement. The project will reduce loading to the nearby wetland. It will also improve the base flow, which is significant to the health of this sensitive receiving water.

Green Streets

Melbourne's Green Your Laneway program is helping to transform the city's laneways into leafy, green, and usable spaces for everyone to enjoy. The city developed an interactive map to show laneways that could go green, based on the amount of sunlight they receive, their exposure to wind, and their physical characteristics. Following a community voting process, it constructed four such laneways, providing the community with multiple environmental, economic, and social benefits.

Green Parking Space

The City of Minneapolis, Minnesota is incentivizing the management of stormwater in parking lots. The city's Stormwater Utility Fee pays for maintenance of the stormwater system and helps prevent and correct stormwater runoff problems. The fee is monthly and is based on impervious area, including parking lots. The Stormwater Credits Program offers fee reductions to owners who use approved methods to manage stormwater runoff on their properties. Fees can also be reduced through replacement of impervious areas with green space.

Underground Storage

Rotterdam's underground water storage solution temporarily collects rainwater in heavy/prolonged showers, reducing combined sewer overflows. The system is incorporated into the design of parking garages, with their two functions – parking and collecting water – always remaining separate from each other. The system temporarily absorbs excess rainwater. As soon as the rain stops, the underground water storage solution is pumped empty into the sewer, ready for the next shower.

Green Parks and Urban Forests

A series of best practices have been identified for regions utilizing green parks and urban forests in the management of excess stormwater.

Green Parks

The Kallang River @ Bishan-Ang Mo Kio Park project is a collaboration between Singapore's PUB and National Parks Board that has turned a concrete canal into a picturesque river rich in biodiversity. A combination of natural features and civil engineering techniques were used to give a natural appearance and prevent soil erosion. Students and teachers created an interactive learning trail highlighting the features and rich biodiversity of the site. A preschool learning trail was developed to allow children to learn about and appreciate Singapore's water sources.

Urban Forests

The City of Vancouver's Urban Forest Strategy aims to protect, plant, and manage trees in order to create a diverse, resilient, and beautiful urban forest on public and private lands across the city. The strategy aims to protect the urban forest during development, manage trees for health and safety, engage citizens in the urban forest, monitor the status and condition of the urban forest, and increase the urban forest canopy. The city's urban forest will play a significant role in its green infrastructure strategy of managing stormwater on both public and private land while at the same time improving water quality, enhancing resilience, and improving natural and urban ecosystems.

Water Bodies

A series of best practices have been identified for regions implementing green infrastructure strategies in order to decrease the risk of flood or improve water quality while protecting, conserving, and enhancing the natural environment.

Retention Ponds

The City of London, Ontario has constructed stormwater management ponds to collect rainfall and surface water runoff and enhance water quality. To ensure the ponds function as designed, the city encourages community groups, residents, and local businesses to adopt their local pond and maintain, beautify, enhance, and safeguard it for all to enjoy. The city provides the materials and support needed to complete twice-annual clean-up activities.

Riparian Buffers

Auckland City Council in New Zealand is providing grant funding for landowners, farmers and lifestyle block owners, and community and environmental groups to protect, enhance, and restore riparian margins. To receive funding, projects need to detail how they will improve water quality, the extent of community involvement in the project, how partnerships will be developed, and how knowledge will be shared among participants.

Green Infrastructure and River Engineering

The Riverlands project at Holnicote in the United Kingdom will use earth-moving equipment to allow natural flow, sediment, and biological processes to develop a fully connected stream–wetland system. Habitat restoration will be fast-tracked to help develop more hydrological and ecological diversity on the site. If the project is successful, the same methods will be applied across a wider area of the river.

Floodplain Restoration

The US Federal Emergency Management Agency (FEMA) funds a Residential Property Voluntary Flood Buyouts program. However, FEMA does not buy houses directly from property owners; instead, state and local communities work together to identify areas where buyouts make the most sense. The community then buys the private property, acquires title to it, and clears it. The property must subsequently remain forever open space for public parks, wildlife refuges, and so forth.

Wetland Restoration

The Amsterdam Wetland Project aims to transform the fragmented delta region, which is managed by various site managers, into one large contiguous nature area with well-coordinated water and nature management and recreational opportunities. The project aims to make natural areas accessible, maintain a healthy peat landscape with regional identity, reduce subsidence and carbon dioxide emissions, and protect unique habitats.

Constructed Wetlands

The South Los Angeles Wetlands Park is a constructed wetland that treats stormwater runoff from a local watershed. Located on vacant land in a densely populated community, the constructed wetland has become a functional and attractive part of the landscape. Other features include solar-powered LED lights, observation decks, picnic areas, a natural rock garden, and educational signage.

Green Corridors

The City of Essen in Germany has developed a green main route network with cycling and hiking paths. The project links together areas of greenery and water and demonstrates the multifunctionality of decentralized rainwater management systems for improving water balance, enhancing flood protection, and improving the urban climate. The project ensures that nearly all residents of the city can access a green public area.

Agriculture and Forestry

A series of best practices have been identified for regions implementing natural water retention measures on agricultural land and in forests.

Meadows

The restoration and agricultural use of the floodplain in Riedstadt, Germany involved the restoration of meadows in both the functional and ancient parts of the floodplain. To establish agricultural use of the restored meadows, local farmers were included from the beginning. They were invited to a roadshow about the project, as well as meetings and excursions. A public relations campaign involving a variety of communication tools, including TV, the Internet, excursions, and lectures, was mounted to inform the general public on why and how the project was being undertaken, with emphasis on the multiple benefits it provided.

Soil and Water Quality

New York State is helping farmers protect soil and water quality. It is providing funding to allow them to implement best management practices that reduce greenhouse gas emissions, increase carbon storage in soils, and improve the health and resiliency of farms.

Forest Riparian Buffers

Pennsylvania's Department of Conservation and Natural Resources offered grants for the installation of multifunctional riparian forest buffers throughout the state. Eligible applicants included government agencies, non-government organizations, private businesses, and Pennsylvania residents.

Floodplain Forests

Dynamic LIFE Lines Danube Project 2019–2023 is an Austrian-Slovak initiative that will restore floodplains along the Danube River. Side arms will be increasingly connected to the Danube, riparian forests will be ecologically upgraded, and riverbanks will be renaturized. The project will restore ecosystem functions and services and improve the productivity of riverine and alluvial ecosystems. Best practice methodologies will be developed concerning the removal of invasive alien species, securing of genetic sources of native tree species, and restoration of native alluvial forests.

Extended Case Studies

From the extended case studies of City West Water in Melbourne, New York City's DEP, the State of Washington, and Welsh Water, a series of best practices have been identified regarding the application of demand management and green infrastructure in order to achieve regional water security for humans while at the same time reducing environmental degradation.

Water Allocation

The following water allocation best practices have been identified.

Water Pricing

With Melbourne experiencing lower water storage levels, the Victorian Desalination Plant increases the resilience of the city's water supply. City West Water's water prices for domestic and nondomestic users will increase to cover the cost of the water ordered from the Desalination Plant.

Water Trading

In the State of Washington, water belongs to the public and cannot be owned by any individual or group. Instead, a person or group may be granted a right to use a volume of water, for a defined purpose, in a specific place. The state follows the "first in time, first in right" rule. If a water shortage occurs, senior rights are satisfied first, and the junior right holders can be curtailed. Water banking allows the transfer of existing water rights from willing sellers to buyers who will put the water to a new use.

Water Augmentation

The following water augmentation best practices have been identified.

Source Water Protection

DEP's Bluebelt Program preserves natural drainage corridors, including streams, ponds, and wetlands, and enhances them to manage stormwater runoff. There is an ongoing program to purchase wetland properties for inclusion in the bluebelt system.

The State of Washington provides financial assistance to the water system and local government projects that protect public drinking water sources. Projects can benefit either water quality or water quantity, or both. Funded activities must contribute to better protecting one or more public water supply sources.

Water Efficiency

The following water efficiency best practices have been identified.

Water Metering

City West Water is cooperating with other water utilities across Melbourne on how and when to roll out digital meters based on customer benefits. To maximize benefits to customers, the utilities are conducting research, trialing technologies, and assessing the financial implications of digital metering in order to determine whether they are viable and deliver benefits to customers, the community, and utilities.

City West Water's digital metering trial aims to determine whether digital water meters can significantly reduce water loss in the network by providing more timely alerts to both City West Water and its customers. There is no cost to customers who participate in the trial.

DEP's automated meter reading system sends daily readings to a computerized billing system up to four times a day. The technology allows customers to view and manage their water consumption on a daily, weekly, monthly, and yearly basis by logging in to their My DEP Account.

Residential and Commercial Water Efficiency

City West Water's residential customers are offered a range of water-efficient toilets for purchase. Its shower head exchange program lets residential customers swap their old shower heads for water-efficient ones for free.

DEP provides funding for water demand reduction projects in city-owned facilities, including those owned by the Department of Education, the Department of Parks and Recreation, the New York City Fire Department, and the City University of New York.

Welsh Water has partnered with a private-sector organization to provide customers with free water-saving gadgets for use around the house.

Industrial Water Efficiency

City West Water provides a free online benchmarking tool that allows business customers to measure their water use against similar companies from the same industry and improve their consumption. Data are collected from water users from around Australia, with water efficiency benchmarks created for many industries.

Welsh Water helps all business sectors make the most efficient and cost-effective use of water by offering a range of services, including water efficiency audits to detect hidden leaks, water efficiency surveys to assess domestic water use on business sites, process audits to assess water consumption within production and operational processes, and telemetry services to help business customers understand their water consumption patterns.

Agricultural Water Efficiency

The State of Washington offers a grant to help landowners install irrigation systems that save water and enhance crop production while protecting endangered aquatic species. The grant funds the deployment of technicians to assess potential water savings projects, along with the design, development, and administration of such projects. The technicians also produce irrigation water management plans to ensure the wise use of water.

Water Reuse and Water Recycling

The following water reuse and water recycling best practices have been identified.

Stormwater Harvesting

City West Water provides funding for councils and private entities in the utility's service area to implement local stormwater harvesting projects that keep public parks, gardens, and sports fields green while enhancing livability in the community. Councils and private entities can apply either singly or in groups.

Water Recycling

City West Water's recycled water is provided for use around the home and garden, such as for watering gardens, washing cars, doing laundry, and flushing toilets. It is only available in estates where purple-colored faucets and pipes have been installed.

On-site Nonpotable Reuse Systems

DEP provides commercial, industrial, and multifamily residential property owners with incentives to install on-site water reuse systems. Customers may be eligible to participate in the program if they have an advanced water meter, and the project can save significant amounts of water per year.

Green Buildings and Green Streets

The following green buildings and green streets best practices have been identified.

Rain Gardens

DEP provides a grant for private landowners to design and construct green infrastructure systems, including green roofs and infiltration projects such as rain gardens and porous pavement. The green roof projects undergo two advanced design review processes in order to be eligible for funding.

Bioswales

Welsh Water is developing and using new, innovative green infrastructure solutions to manage the amount of surface water entering sewers, including bioswales, porous paving, filter strips, grass channels, and geocellular storage. These solutions will help reduce the risk of sewer flooding and pollution and increase sewer capacity. The initiative will also support economic development and protect customers against climate change.

Water Bodies

The following water bodies best practices have been identified.

Floodplain Restoration

The State of Washington has developed a public–private partnership initiative to accelerate integrated efforts to reduce flood risks and restore habitat along the state's major river corridors. The goal is to increase the resilience of floodplains in order to protect local communities and the health of the environment. The program provides grants for collaborative and innovative projects that support the integration of flood hazard reduction with ecological preservation and restoration. The projects can also support other community needs, such as agriculture preservation, water quality improvements, and increased recreational opportunities.

Risk Management Authorities and community groups can work with Welsh Water to manage the risk from surface water flooding. Welsh Water will consider contributing to surface water flooding alleviation projects that reduce the risk of flooding to vulnerable communities, reduce peak river flows, reduce surface water entering Welsh Water's networks in affected catchments, and reduce the impact of flood events on water quality. The utility will contribute to solutions that encourage a partnership approach to the management of surface water flooding using natural flood management or green infrastructure solutions.

Agriculture and Forestry

The following agricultural and forestry best practice has been identified.

Soil and Water Quality

The State of Washington's voluntary agricultural conservation easements give landowners the option to sell their right to develop their agricultural property to another entity. The easements are permanent restrictions on the development and subdivision of a property title, and provide a plan to protect the conservation values of the land. The tool pays landowners the portion of the property value that the development right represents, while leaving them all other property rights.

13

Conclusions

Regional water security around the world is at risk from climatic and nonclimatic challenges, including extreme weather events, rapid population growth and urbanization, economic growth and rising income levels, aging infrastructure, and increased demand for energy and food, impacting water quantity and water quality. Both demand management and green infrastructure have a crucial role in managing water quantity and water quality to achieve water security, where demand management seeks to encourage the better use of existing water supplies through economic and efficient management before further increasing the supply, while green infrastructure involves the use of natural or semi-natural systems that utilize nature's ecosystem services in the management of water resources and associated risks.

From case studies of regions implementing a range of policy innovations that promote the application of demand management and green infrastructure in managing water resources to achieve water security, a range of best practices have been identified in the demand management-related areas of water allocation, water augmentation, water efficiency, and water reuse and recycling, as well as in the application of green infrastructure at the building and street level and in parks and urban forests, water bodies, and on agricultural and forestry lands. These best practices can be implemented by other regions attempting to achieve regional water security for humans while protecting and restoring the natural environment.

Demand Management

Water Allocation

Water allocation is the process whereby an available water resource is distributed (or redistributed) to achieve a series of key policy objectives, including equity, environmental protection, economic and social development, avoidance of frequent or unexpected water deficits, and achievement of the most efficient use of available water. The main water allocation strategies are water pricing, subsidies, water trading, basin-level modeling and seasonal forecasting, and water restrictions. Specifically, the pricing of water and water-related services promotes the efficient use of water and enables its allocation

Regional Water Security, First Edition. Robert C. Brears.

among competing uses. Subsidies are widely used by governments to encourage more environmentally beneficial behavior (e.g., purchasing of water-efficient technologies). Water markets enable limited water resources to be put to their most productive use by distributing them to those who value them most highly over the short and long term. Basin-level modeling helps optimize water allocation among competing water uses and users in response to changing water availability due to seasonal and other changes. Finally, temporary and permanent water restrictions aim to mitigate water source depletion by restricting activities and embedding water savings in new and retrofit builds.

A series of best practices have been identified from regions around the world reforming their water allocation regimes to manage the risk of shortage and adjudicate between competing users. Water price increases can be phased in to meet rising infrastructure costs and ensure the wise use of water. Subsidies can be provided to nondomestic customers to help reduce their water use, which in turn lowers operational costs; examples are subsidies for businesses that implement permanent water savings and discounted water rates for manufacturers that implement water conservation initiatives. Water trading markets can be developed that allow for water rights to be traded either permanently or temporarily; this can increase productivity, allow choice and flexibility in business decision-making, and reduce risk. With regard to basin-level modeling and seasonal forecasting, water models can be developed that enable users to predict how climate, land-use, and other environmental changes will affect the aquatic ecosystems and water resources available for use by businesses and communities. On the topic of water restrictions, cities can prohibit water waste, restrict outdoor watering over spring/summer, and require property owners to fix leaks in sprinkler systems. Prohibitions can be enforced through written warnings, with fines for continued violations.

Water Augmentation

Water augmentation strategies aim to increase the available supply of water. The main ones are the conjunctive use and development of surface and groundwater, managed aquifer recharge (MAR), flood MAR, and the protection of water recharge areas. Specifically, the conjunctive use and development of surface and groundwater involves the recharge of groundwater while surface water is maximized during wet periods and the recovery of water from groundwater resources during dry periods. MAR involves recharging aquifers using either surface or underground recharge techniques with the stored water used when surface supplies are low. Flood MAR uses flood water resulting from, or in anticipation of, rainfall or snowmelt for MAR on agricultural lands and working landscapes. Source water protection involves implementing programs or activities targeted at reducing the likelihood of contaminants polluting drinking water resources and reducing overuse of these resources.

A series of best practices have been identified from regions around the world implementing water augmentation strategies to achieve regional water security. Groundwater storage programs can be implemented to improve regional water supply reliability with the recovery of stored groundwater in dry years and its replenishment in wet years. Cities can develop climate-independent water sources by treating recycled water to a level of purity that is equal to or greater than drinking water quality before injecting it into aquifers for later abstraction for drinking water. Locations can determine the potential for using flood water for MAR with models estimating flood stage, flow, and floodplain inundation, as well as recharge capacities. Source water protection zones can be established, with restrictions easing with increasing distance from the well. To establish new source water protection areas, locations can implement hydrogeological studies, analyze risks to aquifers, inform multiple stakeholders of the process, and provide the public with relevant map materials.

Water Efficiency

Water efficiency strategies aim to reduce water scarcity and maximize the benefits provided by the existing water infrastructure. They also free up water for other users, both human and natural. The main water efficiency strategies are water metering, reducing unaccounted-for water (UFW), public education and awareness, school education, developing a skills base, water labeling, and encouraging water efficiency in buildings, industry, and agriculture. Specifically, water meters facilitate leak detection and energy reduction and enable water customers to compare their consumption against that of others. Reducing UFW can be achieved by improving leakage detection, optimizing pipe maintenance strategies, and developing an instant pressure-regulation system. Public education is crucial in generating an understanding of water scarcity and creating acceptance of the need to promote water efficiency-related programs. Water efficiency programs in schools provide an opportunity for students to regard water conservation as the norm, making it more likely that there will be a lasting change in how water is used across society. Post-secondary education and training on water efficiency provide opportunities for lifelong learning about the environmental gains and economic opportunities associated with water efficiency. Water labeling schemes are essential for reducing water consumption by eliminating unsustainable products from the market. There are multiple avenues to increasing water efficiency in buildings, whether residential, institutional, or commercial. Meanwhile, water efficiency opportunities exist within industrial activities and in agricultural operations.

A series of best practices have been identified from regions around the world promoting water efficiency measures to reduce water scarcity and maximize the benefits provided by existing water infrastructure. Water utilities can offer customers high-usage alerts that send out notifications when water usage is higher than usual. Customers can receive usage alerts and notifications via email, access usage information via computer or phone, view their historical water usage, and track their history. To reduce UFW, locations can utilize satellites to scan for water leaks underground and dispatch field crews to repair them. There are a variety of tools available to help domestic water users conserve water, including online videos that allow homeowners to transform their properties into water-efficient landscapes. Governments at all levels can implement school education programs that help schools conserve water and other resources and integrate sustainability into the curriculum, while at the same time offering best practice guides and professional learning opportunities in water efficiency for teachers and facility managers. Water utilities can run water efficiency training programs to help business customers lower their water consumption and operating costs while protecting the environment. Water labeling schemes can be developed to help consumers compare the water efficiencies of different products and determine a given product's rate of water consumption. Cities can implement water and energy use benchmarking ordinances to diagnose energy-inefficient buildings and aid in the implementation of energy-saving features, with schemes involving energy and water audits, retro-commissioning, and energy and water reduction targets. Water utilities can offer industrial water efficiency services to help businesses lower their water and energy bills while reducing carbon emissions. Examples of services include process audits, on-site inspections, and benchmarking services. Finally, on-farm irrigation efficiency can be enhanced through funding aimed at improving infrastructure and returning saved water to the environment.

Water Reuse and Recycling

Water reuse is defined as the reuse of water within a single process or use of harvested water for another purpose, without treatment. In contrast, water recycling is the use of harvested water for another purpose, after treatment. The main water reuse and water recycling strategies are rainwater harvesting, stormwater harvesting, water recycling, potable reuse, on-site nonpotable reuse, environmental flows, and the development of alternative water sources. Specifically, rainwater harvesting involves the collection of precipitation running off roof surfaces into cisterns or other storage vessels. Stormwater harvesting involves collecting, storing, and treating stormwater from urban areas, which can then be reused for a variety of nonpotable uses. Recycled water can be used for a range of purposes, including in urban environments, industry, and agriculture. Water recycling for potable reuse involves the planned use of treated municipal wastewater for augmenting drinking water supplies. Indirect potable reuse systems involve communities using surface water or groundwater containing treated wastewater as a drinking source, while direct potable reuse systems introduce purified water directly back into an existing water supply system. Onsite nonpotable water systems capture and treat water sources generated from within or around a building, such as roof runoff, stormwater, domestic wastewater, or graywater. Treated wastewater can be used to restore and enhance natural habitats such as wetlands and marshes or to maintain flows of small water bodies, which may preserve or restore biodiversity. Meanwhile, a variety of alternative water sources can be developed, including capturing condensate, generating water from the atmosphere, reusing water from industrial water purification systems, capturing and reusing foundation and blowdown water, desalination, and fog harvesting.

A series of best practices have been identified from regions around the world implementing a range of water reuse and water recycling technologies. Water utilities can provide rebates to customers for installing rainwater harvesting systems, with eligibility subject to keeping the system in service for a set time, agreeing to pre- and post-installation inspections by the utility, and being willing to let the utility reference the project in any outreach materials. City park agencies can install stormwater harvesting systems to help keep gardens healthy in a changing climate, which in addition to saving drinking water reduces pollutants entering waterways. Funding can be made available for high-impact and innovative projects that treat and reclaim industrial used water for process reuse. Both indirect and direct potable reuse systems enable cities to become more drought-tolerant by extending potable supplies. Cities can offer grant assistance for the construction of nonpotable reuse systems at both the building and district scale, with discounts on water and wastewater fees offered to successful systems. In addition to a variety of nonpotable uses in buildings, recycled water can also be used for the restoration of waterways, which has the additional benefit of providing space for recreation and relaxation. Regions can also explore a variety of alternative water sources, including capturing condensate, installing atmospheric water generators, developing renewable energy-powered desalination plants, and installing fog harvesting systems in rural communities.

Green Infrastructure

Building and Street Level

Green infrastructure can be applied at the building and street levels and includes a wide range of water features designed to manage stormwater runoff and improve water quality. At the building level, it includes green and blue roofs, rain barrels and cisterns, rain gardens, and bioswales, as well as permeable pavement. Specifically, green roofs are highly engineered systems comprising plantings, drainage layers, and watertight membranes that reduce runoff and runoff peak attenuation. Blue roofs provide temporary storage and slow release of rainwater on a rooftop. Rain barrels and cisterns reduce runoff volume and, during storms, delay and reduce peak runoff flow rates. Rain barrels are usually used for individual residences, while cisterns have both residential and commercial applications. Rain gardens are designed to receive water runoff from a roof, driveway, or lawn. They slow stormwater flow, help remove pollutants, and recharge freshwater bodies, all the while looking attractive. Bioswales can be incorporated into the landscape to slow runoff while removing pollutants. There are a variety of permeable surfaces available to reduce stormwater runoff flows and improve water quality, including permeable pavement, pervious concrete, porous asphalt, and permeable pavers. At the street level, green streets use a variety of systems, including vegetated curb extensions, sidewalk planters, landscaped medians, vegetated swales, permeable pavement, and street trees, to reduce stormwater, improve water quality, and enhance pedestrian safety in residential, commercial, and arterial streets and alleys. Green parking lots use green infrastructure that closely mimics the natural water cycle to manage stormwater through rainfall retention, pollutant removal, and water infiltration. Finally, on-site, underground storage infrastructure can capture and store stormwater collected from surrounding impervious areas.

A series of best practices have been identified from regions around the world implementing green infrastructure at the building and street levels. Municipalities can provide subsidies to homeowners, companies, associations, and foundations for the construction of green roofs, with the amount provided based on water storage capacity and green area. Cities can also use subsidies to encourage the development of multifunctional green roofs. Municipality agencies can partner with retail stores to provide free rain barrels and offer discounted cisterns to homeowners, schools, and businesses. They can construct rain gardens in neighborhoods following a design and construction process that includes site selection, engaging with impacted property owners, and maintenance. They can implement green infrastructure projects that increase access to green space, enhance rainwater management, and improve neighborhood amenities. In derelict areas, they can convert residential streets into green infrastructure by using pervious pavement to manage excess stormwater and improve the health of surrounding water bodies. Municipalities can transform their alleyways and lanes into green spaces that provide the community with multiple environmental, economic, and social benefits. City agencies can offer stormwater credits to parking lot owners who use approved methods to manage stormwater runoff. Cities with limited space can implement underground water storage solutions to temporarily collect rainwater in heavy/prolonged showers, reducing combined sewer overflows (e.g., a storage system incorporated into the design of a parking garage).

Parks and Urban Forests

Parks and urban forests can intercept and filter stormwater runoff, preventing flooding and improving water quality. Specifically, parks can capture rainfall where it falls. Excess water from surrounding areas can also be directed into parks and stored there for later uses in dry periods. This reduces the volume of stormwater requiring treatment in wastewater treatment plants and reduces pollution of water bodies. There are two main methods for managing stormwater in parks: holding it for conventional treatment and using natural processes to percolate it through the soil. Park infrastructure can be built or retrofitted to treat stormwater or drain it to enhanced pervious surfaces. Many green infrastructure best practices can be implemented to manage stormwater in parks, including infiltration beds, rain gardens, stormwater planter boxes, porous pavement, green and blue roofs, and constructed wetlands. Urban forests are networks or systems comprising all woodlands, groups of trees, and individual trees located in urban and peri-urban areas. They include forests, street trees, trees in parks and gardens, and trees in derelict locations. Urban and peri-urban forestry is the practice of managing urban forests to ensure their optimal contribution to the physiological, sociological, and economic well-being of urban societies. Well-managed and healthy urban forests can contribute to the sustainable management of water and water resources in urban and peri-urban watersheds, helping to clean, save, and store water and reduce the risk of water-related disasters.

A series of best practices have been identified from regions around the world utilizing green parks and urban forests in the management of excess stormwater. Water utilities can collaborate with municipal agencies to turn concrete canals running through parks into natural water bodies that manage excess water while offering learning opportunities for visitors. Cities can develop urban forest strategies that aim to protect, plant, and manage trees in order to create a diverse, resilient, and beautiful urban forest on both public and private lands, with multiple co-benefits including managing stormwater, improving water quality, enhancing resilience, and improving natural and urban ecosystems.

Water Bodies

Local and regional authorities are generally responsible for land-use decisions and have an essential role in assessing environmental impacts and protecting, conserving, and enhancing natural ecosystems, including water bodies. Incorporating green infrastructure that decreases flood risk or improves water quality into related plans and strategies can overcome the fragmentation of habitats and preserve or restore ecological connectivity and enhance ecosystem resilience, all the while providing multiple social benefits. A variety of green infrastructure strategies can be implemented, including retention ponds and riparian buffers, river engineering, floodplain restoration, wetland restoration, and green corridor initiatives. Specifically, retention ponds are designed to capture and slowly release stormwater over a period of time. They also create wildlife and aquatic habitat and provide recreational, esthetic, and open space opportunities. Riparian buffers are strips of vegetation established next to waterways in managed landscapes that are designed to capture runoff, nutrients, and sediment while restoring the natural aquatic environment. Green infrastructure can promote the conservation or restoration of the natural character of rivers, with two types of measure available: green measures that use vegetative materials exclusively and green-gray measures that use a mix of green and gray nonvegetative, nonbiodegradable components to support structural stability. Floodplain restoration reestablishes the structure and function of ecosystems and floodplains, returning them to their natural conditions and functions prior to development. Wetland restoration involves enhancing wetland storage capacity by damming the outflow route to increase available overland water storage capacity or by undertaking floodplain drainage to increase soil water storage capacity. Other wetland restoration activities include keeping stock out, controlling weeds and plants, and planting. Green corridors are natural routes for the movement of people, animals, seeds, and water. They can be used to link trees, parks, green spaces, and drainage areas with vegetated pedestrian walkways, cycle paths, and transport routes to increase biodiversity while building resilience to floods.

A series of best practices have been identified from regions around the world implementing green infrastructure strategies to decrease flood risk or improve water quality while protecting, conserving, and enhancing the natural environment, including water bodies. Municipalities can construct stormwater management ponds to collect rainfall and surface water runoff while enhancing water quality. Ponds can be adopted by local community groups and businesses who volunteer to maintain, beautify, enhance, and safeguard them for all to enjoy, with city agencies providing materials and support to complete clean-up activities. Local governments can offer grant funding to farmers and community and environmental groups to restore riparian buffers, with the amount based on the project's ability to improve water quality, promote community engagement, and foster partnerships and knowledge sharing among project participants. River systems can be restored to their natural processes and habitats using earthmoving equipment, allowing natural flow, sediment, and biological processes to develop fully connected stream–wetland systems. Government agencies can restore floodplains through voluntary flood buyout programs, providing open space for public parks or wildlife refuges. Regional wetland restoration projects can be implemented to transform fragmented areas into contiguous natural spaces with well-coordinated water and nature management and recreational opportunities. Cities can develop constructed wetlands on vacant sites and brownfields to treat stormwater runoff from local watersheds. These can be multifunctional, providing walking trails, picnic areas, and educational signage. Cities can also develop multifunctional green corridors that in addition to providing residents with a network of cycling and hiking paths, enhance flood protection and improve the urban climate.

Agricultural and Forestry Lands

Natural water retention measures (NWRM) are a form of green infrastructure that can play an essential role in catchment-scale flood risk management. NWRM on agricultural land and in forests can help reduce the risk of downstream flooding by enhancing or restoring natural hydrological processes, including interception, evapotranspiration, infiltration, and ponding, in addition to improving water quality. It can take the form of meadows, buffer strips, crop rotation, conservation tillage, cover crops, and mulch, as well as forest riparian buffers and floodplain forests. Specifically, meadows can be restored to increase the uptake and storage of water during temporary floods. Buffer strips can slow runoff and absorb excess water, while at the same time reducing the volume of suspended solids, nitrates, and phosphates originating from agricultural runoff. Crop rotation can maintain or improve soil organic matter content, which in turn reduces erosion and increases infiltration capacity, reducing downstream flood risk. Conservation tillage can slow water movement, which reduces the amount of soil erosion and potentially leads to greater infiltration. Cover crops improve organic matter levels and increase soil water storage, reducing downstream flood risk. Mulching adds organic matter to the soil, reduces weed growth, and reduces erosion during the period when the ground is covered in the mulch. Forest riparian buffers provide flood control, with trees diffusing the energy of floodwaters, reducing damage downstream. Meanwhile, forests in floodplains can increase their hydraulic roughness, slowing the release of stored surface water.

A series of best practices have been identified from regions around the world implementing NWRM on agricultural land and in forests. Local governments can restore meadows in both functional and ancient floodplains, with projects involving farmers at all stages. Public relations campaigns, involving a variety of communication tools including TV, the Internet, excursions, and lectures, can inform the public on why and how restoration projects are undertaken, with emphasis placed on their multiple benefits. Governments can help farmers protect soil and water quality by providing funding to implement best management practices that reduce flooding risks, increase carbon storage in soils, and improve the health and resiliency of farms. Authorities can provide grants for the installation of multifunctional riparian forest buffers. Large-scale floodplain restoration activities can be undertaken that involve connecting side arms of rivers, upgrading riparian forests, and renaturalizing riverbanks.

Implementing Best Practices in Other Regions

From extended case studies of regions implementing demand management and green infrastructure to achieve water security, a series of best practices have been identified for other regions around the world attempting to achieve regional water security for both humans and nature. Regarding water allocation, cities that are experiencing warmer, drier weather conditions with less rain will likely rely more on desalination to enhance the resilience of their water supply. As such, water prices for domestic and nondomestic users will need to increase to cover the cost of the water ordered from the desalination plant. Water rights trading systems allow the transfer of existing water rights from willing sellers to buyers who want to put the water to a new use. This ensures water is moved to where it is needed the most. On the matter of water augmentation and source water protection, grants can be provided to water systems and local government projects to protect public drinking water sources (whether in terms of water quality or water quantity, or both). Regarding water efficiency, water utilities can cooperate on how best to roll out digital water meters in order to maximize benefits to customers. Meanwhile, the use of automatic meter reading technology enables customers to view and manage their water consumption on a daily, weekly, monthly, and yearly basis through an Internet portal. In residential buildings, water utilities can offer a range of water-efficient toilets for sale, as well as providing a showerhead exchange program that lets customers swap their old shower heads for new water-efficient ones for free. They can also provide funding for water demand reduction projects in city-owned facilities. For industrial customers, water utilities can provide free online benchmarking tools that allow them to measure their water use against similar companies from the same industry and improve their consumption. Utilities can also provide process water audits to help large water-using customers assess water consumption within production and operational processes. Finally, grants can be provided to help agricultural producers install irrigation systems that save water and enhance crop production while protecting endangered aquatic species. With regard to water reuse and water recycling, utilities can provide funding to local governments and private entities to implement stormwater harvesting projects for nonpotable uses, including irrigation of sporting fields. Recycled water can also be provided to domestic customers for nonpotable uses. Meanwhile, municipalities can provide funding for large customers to install on-site water reuse systems for buildings. On the application of green infrastructure in the context of green buildings and green streets, water managers can provide grants for private landowners to design and construct green infrastructure systems including green roofs and infiltration projects such as rain gardens and porous pavement. Water utilities can also develop and use new innovative solutions to manage the amount of surface water entering the sewers, helping reduce the risk of sewer flooding and pollution. In the context of green infrastructure in water bodies, public–private partnerships can be developed to reduce flood risks and restore habitats along major river corridors, with grants offered to facilitate the development of collaborative and innovative projects that support the integration of flood hazard reduction with ecological preservation and restoration. Meanwhile, government authorities and community groups can work with utilities to manage the risk from surface water flooding while at the same time reducing the impact of flood events on water quality using green infrastructure solutions. In terms of green infrastructure on agricultural and forestry lands, agricultural conservation easements can be developed to protect the conservation values of land, with landowners paid the portion of the property value that the development right represents while retaining all other property rights.

In conclusion, water authorities at multiple levels of governance in differing regions of the world can implement a range of policy innovations that promote the application of demand management and green infrastructure to achieve regional water security for humans while protecting and restoring the natural environment.

Index

y